Inertial Sensor Assessment of Human Movement

Inertial Sensor Assessment of Human Movement

Guest Editors

Elissavet Rousanoglou

John Buckley

Alan Godfrey

MDPI

Basel • Beijing • Wuhan • Barcelona • Belgrade • Novi Sad • Cluj • Manchester

Guest Editors
Elissavet Rousanoglou
Department of Sports
Medicine and Biology
of Exercise
National and Kapodistrian
University of Athens
Athens
Greece

John Buckley
School of Computing
& Engineering
University of Bradford
Bradford
UK

Alan Godfrey
Department
Computer Science
Northumbria University
Newcastle-upon-Tyne
UK

Editorial Office
MDPI AG
Grosspeteranlage 5
4052 Basel, Switzerland

This is a reprint of the Special Issue, published open access by the journal *Biomechanics* (ISSN 2673-7078), freely accessible at: https://www.mdpi.com/journal/biomechanics/special_issues/2K99XM88N3.

For citation purposes, cite each article independently as indicated on the article page online and as indicated below:

Lastname, A.A.; Lastname, B.B. Article Title. *Journal Name* **Year**, *Volume Number*, Page Range.

ISBN 978-3-7258-6934-3 (Hbk)
ISBN 978-3-7258-6935-0 (PDF)
https://doi.org/10.3390/books978-3-7258-6935-0

Contents

About the Editors

Elissavet Rousanoglou

Elissavet Rousanoglou is a Full Professor in Sport Biomechanics at the National and Kapodistrian University of Athens, Greece, and the Director of the Sport Biomechanics Laboratory for the School of Physical Education and Sport Science. As a member of the Sector of Sport Medicine and Biology of Exercise, she teaches courses in basic and advanced Sport Biomechanics, Experimental Biomechanics, Applied Sport Biomechanics, Kinesiology, Ergonomics, Analysis of Kinesio-Biological Signals, and Acoustic Stimulus and Movement Performance.

She completed her undergraduate and postgraduate studies at the National and Kapodistrian University of Athens, School of Physical Education and Sport Science, Greece, and gained research experience in the Human Performance Lab at the University of Calgary, Canada.

She is a founding member of the Hellenic (Greek) Society of Biomechanics (HSB), has been its General Secretary for many years, and has also served as its President (two years). She is a reviewer for the 2026 World Congress of Biomechanics, as well as a reviewer for numerous papers in a variety of biomechanics and motor control-related journals. She has been a member of the scientific and organizing committees of national and international biomechanics congresses.

She has authored or co-authored about 50 scientific papers and more than 100 oral and poster presentations at scientific congresses, has been a Guest Editor for Special Issues in scientific journals, and also edited the Greek version of the academic textbooks: "*Fundamental Basis of Human Movement*"—3rd and 5th Editions; "*Biomechanics of Sport and Exercise*"—4th Edition; "*Rhythm, Music and the Brain*"—1st Edition; and "*Methods of Group Exercise Instruction*"—3rd and 4th Editions.

She is a member of the Biomechanics Initiative Team and is very enthusiastic about advocating for the yearly worldwide event of National Biomechanics Day to young students.

John Buckley

John Buckley is a Reader in Movement Biomechanics at the School of Computing and Engineering at the University of Bradford (UK). He teaches courses in clinical movement analysis, biomechanics, and rehabilitation engineering, and he conducts interdisciplinary research in the areas of clinical biomechanics and human movement. His research interests are in developing and using motion analysis approaches to determine the movement adaptations and/or compensatory mechanisms used for locomotion by individuals with musculoskeletal abnormalities, problems or dysfunction, and/or by those with sensory impairment. He has published over 90 academic papers in both rehabilitation-specific journals and multi-disciplinary journals such as *Experimental Gerontology, Experimental Brain Research* and *PLoS One*. Dr Buckley is a member of the International Society for Posture and Gait Research and a Fellow of The Higher Education Academy.

Alan Godfrey

Alan Godfrey is an Associate Professor and previously led his school's Digital Health & Wellbeing research group. He received his BEng in electronic engineering and PhD in biomedical electronics from the University of Limerick, Ireland. He also holds an MBA from Newcastle University (2017).

His major research interests are in wearable technology and algorithms for data science and analytics in healthcare. This includes areas of artificial intelligence (AI), machine learning, data mining and multidimensional signal processing. He has published well over 200 papers on these topics in various engineering and medical journals from a portfolio of translational-based research. He serves as Deputy Editor for *npj/Nature Digital Medicine*, Editor for *Maturitas*, and Associate Editor for the *Journal of NeuroEngineering & Rehabilitation*. He was previously an International Advisory Board Member for *Physiological Measurement*.

Alan Godfrey is a Member of the Institution of Engineering Technology (MIET) and a Senior Member of the Institution of Electrical and Electronic Engineering (SMIEEE). He is a Fellow of the Higher Education Academy (FHEA).

biomechanics

MDPI

Article

Inertial Sensing of the Abdominal Wall Kinematics during Diaphragmatic Breathing in Head Standing

Elissavet Rousanoglou *, Apostolina Foskolou, Analina Emmanouil and Konstantinos Boudolos

School of Physical Education and Sport Science, National and Kapodistrian University of Athens, 172 37 Daphne, Greece; anemmanouil@phed.uoa.gr (A.E.); cbountol@phed.uoa.gr (K.B.)
* Correspondence: erousan@phed.uoa.gr

Abstract: Head standing (HS) in concurrence with diaphragmatic breathing is an atypical deviation from daily activity, yet commonly practiced. The study aimed at the inertially sensed effect of diaphragmatic versus normal breathing on the abdomen wall kinematics during HS. Twenty-eight men and women maintained HS and erect standing (ES) under normal and diaphragmatic breathing. An inertial sensor (LORD MicroStrain®, 3DM-GX3®-45, 2 cm above the umbilicus, 100 Hz, MicroStrain, Williston, VT, USA) recorded the 3D abdomen wall angular displacement (AD) (bandpass filter (0.1–0.5 Hz)). ANOVAs ($p \leq 0.05$, SPSS 28.0) were applied to the extracted variables (AD path: magnitude, individual variability-%CVind, and diaphragmatic to normal ratio). Reliability measures (ICC and %SEM) and the minimal detectable change (%MDC90) were estimated. Diaphragmatic breathing increased the AD path ($p \leq 0.05$) with the diaphragmatic to normal ratio being lower in HS ($p \leq 0.05$). The similar AD time series (cross-correlations at $p \leq 0.05$) and the ICCs (>0.80) indicated excellent reliability with the similar across conditions %CVind ($p \leq 0.05$), further enhancing reliability. The %MDC90 was consistently higher than the %SEM upper boundary, indicating the differences as "real" ones. The results contribute to the limited data concerning a widely practiced atypical deviation from daily activity, as HS in concurrence with diaphragmatic breathing.

Keywords: inverted stance; reliability; respiration

1. Introduction

Head-down inversions present a particularly atypical deviation from daily activity and upset the various systemic adaptations for overcoming the effects of gravity, a condition that appears to attract an ongoing research interest concerning their potential health benefits [1–5]. The alleged positive health effect of gravity inversion exercises is not a new issue. Many fitness facilities (sports clubs, YMCAs, health spas, etc.) were long ago prompted to install inversion devices so that their members could hang upside down [6]. Among them, most possibly due to the growing popularity of the Eastern ancient tradition of yoga, head standing (HS) not only is known as the "king of all postures" but also is advocated as a cure for almost all health issues by Eastern yoga gurus [7]. Nowadays, many people are encouraged or even challenged not only to practice HS but also to concurrently practice diaphragmatic breathing (characterized by a slow and deep breathing cycle), which is expected to further enhance the alleged HS benefits [7–9].

In diaphragmatic breathing, the inspiration phase aims to increase both the vertical and transverse diameters of the thoracic cavity through the contraction of the diaphragm that depresses its central tendon and elevates the lower ribs, thus pushing the viscera downward and outward [10,11]. During diaphragmatic breathing, the normally passive expiration phase [12] is performed actively to compress the abdomen wall and reverse the viscera's upward and inward phases. As shown in normal breathing, muscles having both a respiratory and a postural role, such as the diaphragm (primary inspiratory muscle) and the abdominals (primary expiratory muscle), and may compromise their breathing function to prioritize their postural role [11,13]. This prioritization could be expected to be more critical in body stances of inherent instability, i.e., HS [14] when an additional postural perturbation is concurrently induced, i.e., diaphragmatic breathing [11,13,15].

During diaphragmatic breathing, in addition to postural stability [14], the breathing mechanics [15] are affected by the influence of the respiratory muscles' activation (i.e., diaphragm-related abdomen wall expansion). Nevertheless, the research-based evidence appears rather limited and focused on the hung upside down position [3–5], where the shoulders and ankles support the body's weight load rather than bearing the whole-body weight load on one's head, i.e., during HS. Studies on the hung head-down position indicate disturbed breathing mechanics due to the reversed action of gravity on the ribcage (*inspiratory direction:* maximal gravity effect superseding the abdominal hydrostatic pressure, increased functional capacity of the lungs, and oxygen utilization nearer to the erect posture level; *expiratory direction:* reversed effect of gravity on the elastic recoil properties of the various contracting components of the respiratory muscles probably appears to have the major role, less postural contraction of the respiratory muscles and, consequently, a smaller functional residual capacity) [5]. In the HS position, the normal expiratory function [13] suggests an increased diaphragmatic restraint because the body weight is borne on the vertex and the rib cage is free of the weight of the shoulder girdle. Considering the abdomen wall expansion in diaphragmatic inspiration as well as the intense role of the abdominal muscles in expiration to breathe out the air volume that was deeply inhaled, the abdomen wall kinematics may provide a quantitative criterion of diaphragmatic versus normal breathing. Despite the technological advancements of non-invasive devices for studying breathing mechanics, i.e., inertial sensors, there appears to be limited information concerning the kinematics of the abdomen wall during diaphragmatic breathing in HS compared to erect standing (ES).

The technology advancements have inspired a lot of recent studies to use inertial sensors (devices that include an accelerometer and a gyro sensor) to detect and evaluate breathing function [16–21]. The use of inertial sensors aimed to overcome the difficulties of the gold standards for testing breathing function (spirometry and body plethysmography) [17] are both rather uncomfortable because the person must breath through a mouthpiece or a face mask with a sealed nose. Nevertheless, most of such studies tested normal breathing in the erect posture. In those studies, accelerometers placed on the chest wall [16,17,19–23] or the abdomen wall [16,18–21,23] appear to effectively detect the breathing kinematic changes in both the chest and the abdomen wall. To the best of our knowledge, a rather small number of studies include deep breathing data using an accelerometer [19,20]. However, the kinematic changes reported by Hung et al. [19] concern deep breathing while standing, sitting, and lying but not in head-down body stances; in addition, the breathing instructions did not target intentional diaphragmatic breathing. Karacocuk et al. [20] also describe their breathing instructions as forceful but quiet, stating that although abdominal (diaphragmatic) rather than thoracic breathing dominated, their sensor location (chest wall) did not allow the detection of abdomen wall kinematic changes. Although the use of accelerometers to assess the kinematic changes in the abdomen or chest wall is not without problems (strong dependency on the sensor location on the participant's body [24] or motion artifacts [20]), such problems appear to be effectively handled through elaborated signal-filtering techniques [17,23,25].

To the best of our knowledge, the studies using inertial sensors and advanced signal-processing techniques concern normal breathing in the erect posture. There appears to be a lack of experimental information concerning the effect of diaphragmatic breathing in concurrence with HS, an extremely atypical deviation from daily activities, nevertheless widely practiced due to its alleged health benefits. We hypothesized that diaphragmatic compared to normal breathing would significantly increase the abdomen wall displacement and that inertial sensing would be feasible to discern this increase.

Thus, the purpose of this study was to apply inertial sensing to evaluate the effect of diaphragmatic versus normal breathing on the abdomen wall kinematics during head standing.

2. Materials and Methods

2.1. Participants

Twenty-eight healthy men and women (32.6 ± 9.7 and 22.4 ± 5.0 years; height: 173.8 ± 10.0 and 166.8 ± 6.0 cm; and body mass: 72.1 ± 9.0 and 59.7 ± 7.2 kg, for the men (n = 8) and the women (n = 20), respectively) were selected among those who volunteered as participants after a public announcement or personal invitation. The inclusion criteria (no musculoskeletal injury during the past 2 months, no history of high blood pressure, no known respiratory or vision disorder, previous experience in HS and diaphragmatic breathing) aimed to ensure that the participants could safely maintain HS in concurrence with diaphragmatic breathing. All participants were fully informed about the purpose of the study, and informed consent was obtained from all of them. Their detailed characteristics are presented in Table 1.

Table 1. Personal characteristics of the participants (N = 28).

a/a	Gender	Age (y)	Body Mass (kg)	Body Height (m)	Body Mass Index (kg/m^2)	* BMI Classification	Headstand History Experience	Headstand Practice Experience	Diaphragmatic Breathing Experience
S01	F	26	63.0	1.61	24.3	Normal	1–2 years	Personal Practice	<6 months
S02	F	25	57.5	1.68	20.4	Normal	>8 years	Artistic Gymnastics	<1 year
S03	M	29	79.0	1.78	24.9	Normal	>8 years	Artistic Gymnastics	<6 months
S04	F	21	55.0	1.60	21.5	Normal	5–7 years	Artistic Gymnastics	<6 months
S05	F	25	51.0	1.67	18.3	Underweight	5–7 years	Artistic Swimming	<1 year
S06	M	27	80.7	1.80	24.9	Normal	<1 years	Personal Practice	<6 months
S07	M	34	68.3	1.67	24.5	Normal	>8 years	Artistic Gymnastics	<6 months
S08	F	33	52.0	1.58	20.8	Normal	1–2 years	Yoga	2–4 years
S09	M	41	72.0	1.82	21.7	Normal	2–4 years	Yoga	>8 years
S10	F	21	54.0	1.67	19.4	Normal	1–2 years	Yoga	1–2 years
S11	F	37	64.0	1.67	22.9	Normal	>8 years	Yoga	>8 years
S12	F	22	72.5	1.72	24.5	Normal	>8 years	Rhythmic Gymnastics	<6 months
S13	F	18	60.0	1.60	23.4	Normal	2–4 years	Yoga	2–4 years
S14	M	51	64.0	1.80	19.8	Normal	5–4 years	Yoga	5–7 years
S15	F	22	65.0	1.70	22.5	Normal	<1 years	Yoga	<1 year
S16	F	20	63.0	1.73	21.0	Normal	2–4 years	Yoga	2–4 years
S17	F	20	54.0	1.64	20.1	Normal	2–4 years	Rhythmic Gymnastics	<6 months
S18	F	19	67.0	1.69	23.5	Normal	>8 years	Rhythmic Gymnastics	<6 months
S19	F	19	62.0	1.72	21.0	Normal	5–7 years	Artistic Gymnastics	<6 months
S20	M	33	72.0	1.70	24.9	Normal	5–7 years	Yoga	>8 years
S21	F	18	66.0	1.79	20.6	Normal	>8 years	Artistic Gymnastics	<6 months
S22	F	20	68.0	1.78	21.5	Normal	>8 years	Rhythmic Gymnastics	2–4 years
S23	F	20	49.5	1.59	19.6	Normal	2–4 years	Artistic Gymnastics	<6 months
S24	M	26	55.0	1.53	23.5	Normal	>8 years	Artistic Gymnastics	>8 years
S25	F	18	57.0	1.62	21.7	Normal	2–4 years	Breakdance	<6 months
S26	F	23	49.7	1.64	18.5	Underweight	>8 years	Artistic Gymnastics	<6 months
S27	F	22	58.4	1.65	21.5	Normal	>8 years	Artistic Gymnastics	<6 months
S28	M	20	76.5	1.80	23.6	Normal	5–7 years	Personal Practice	<6 months

* According to the World Health Organization, https://www.who.int/data/gho/data/themes/topics/topic-details/GHO/body-mass-index, accessed on 10 August 2023.

2.2. Experimental Procedure

2.2.1. Verification of Diaphragmatic Breathing Pattern

Participation clearance was provided only after a familiarization session that aimed to ensure that the volunteers could perform and maintain HS for 40 s and perform the diaphragmatic breathing pattern (Figure 1). Specifically, an experienced examiner evidenced the two characteristic visual alterations of the abdomen wall between regular and diaphragmatic breathing, that is, the intense abdomen wall expansion during the inspiration phase (without observable motion in the upper thorax) as well as its subsequent intense flattening and inward pull during the expiration phase [14,15,26]. In addition to the experienced examiner's visual verification, the diaphragmatic breathing pattern was also experimentally verified through the angular displacement signal of an inertial sensor positioned on the abdomen wall (described below) concerning the abdomen wall rotation around the frontal horizontal axis [20], thus tracing the abdomen wall expansion and compression during inspiration and expiration, respectively. Karacocuk et al. [20] also report that the inclination due to rotation around the frontal horizontal axis (roll angle in the present study) is the most accurate one because it yields the lowest error concerning the breathing rate and its correlation with the respiratory minute volume.

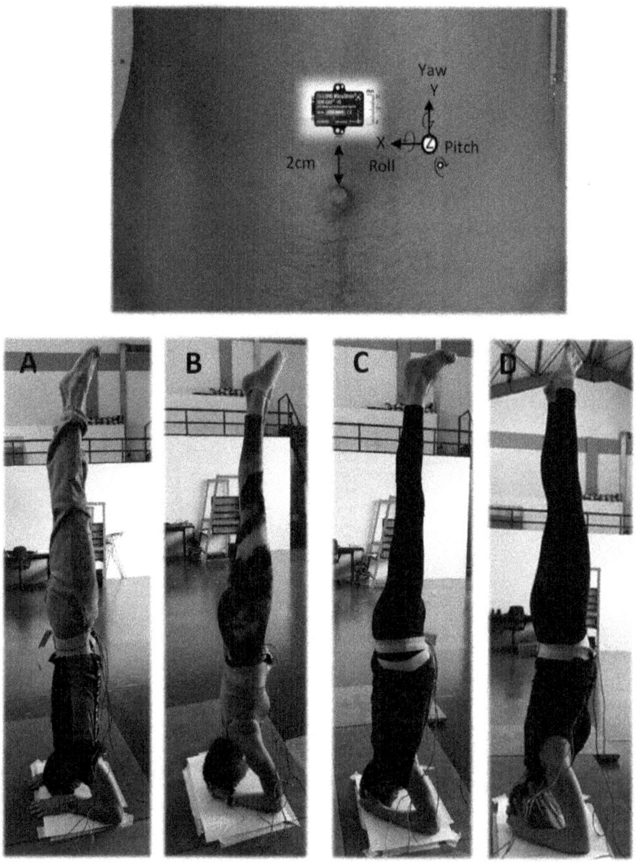

Figure 1. Men and women participants in the head-standing body stance. (**A**,**B**) during normal breathing and (**C**,**D**) during diaphragmatic breathing where the abdomen wall expansion is visible. The anatomical calibration of the sensor is described in the second paragraph of the data collection section.

2.2.2. Data Collection

The participants were instructed to maintain (40 s, barefoot) HS (Figure 1) as well as ES. They maintained both stances under normal and diaphragmatic breathing, both at their preferred breathing frequency. The two body stances were presented in rotation order, aiming to obtain a similar number of participants in each body stance × breathing type combination. HS was maintained without wall support; thus, in all HS trials, a trained assistant was standing next to the participant to safely cushion a potential balance loss. Also, a mat was positioned in the potential falling area in case of balance loss. Three successful trials were obtained per participant in each combination of body stance and breathing type. One minute rest was allowed between trials of the same body stance and 3 min rest before initiating the trials of the subsequent body stance. To eliminate the influence of external factors that could possibly affect the sensor signal, we did our best to maintain the same experimental conditions for all participants (conducting the data collection at fixed times of the day with some tolerance, monitoring air temperature and humidity to keep similar ambient conditions, and ensuring the absence of environmental sound or voice pollutants—a possible reason for a variety of oscillations) [27].

Following previous studies [16,21,23,28], a triaxial inertial sensor was used (LORD MicroStrain®, 3DM-GX3®-45, sampling at 100 Hz, https://www.microstrain.com/sites/default/files/applications/files/3DM-GX3-45-GPS-Aided-Inertial-Navigation-System-Data-Sheet.pdf, accessed on 23 November 2023) enabling the recording of the 3D abdomen wall 3D linear acceleration and 3D angular displacement (AD) during the normal and the diaphragmatic breathing trials. The AD rather than the linear acceleration signal was selected for analysis as a more direct estimate of the abdomen wall displacement. With the participant in ES, the inertial sensor was securely positioned on the abdomen wall, at 2 cm above the umbilicus (Figure 1). Also in ES, the anatomical calibration of the sensor indicated the abdomen wall AD for the inspiration and expiration phases as follows: the roll Euler angle as the forward–backward motion of the abdomen wall in the sagittal plane (rotation around the frontal horizontal axis, ***forward expansion during*** diaphragmatic inspiration); the pitch Euler angle as the outward–inward motion in the frontal plane (rotation around the sagittal horizontal axis, ***outward expansion during*** diaphragmatic inspiration); and the yaw Euler angle as the downward–upward motion in the transverse plane (rotation around the vertical axis, ***downward expansion during*** diaphragmatic inspiration). The AD signal was recorded in radians and converted into degrees (1 radian = 57.3 degrees). Then, the resultant of the roll, pitch, and yaw signals was also calculated.

2.2.3. Data Procession

Initially, all raw Euler angle signals were pre-processed (MatLab R2022b, bandpass filtering (https://www.mathworks.com/help/signal/ref/bandpass.html, accessed on 10 August 2023) by applying the function y = bandpass(x,fpass,fs), where y stands for the filtered signal, fs specifies the sampling frequency (that is 100 Hz in the present study), and fpass specifies the two-element vector concerning the passband frequency range of the filter in Hz. Following previous studies [17] and an extensive trial and error process, 0.1 Hz and 0.5 Hz were decided as the optimum lower and upper cut-off frequencies, respectively, for the passband frequency range.

The pre- and post-filtering signal to noise ratios (SNRs) in each body stance and breathing type condition are presented in Table 2 for all three Euler angles, as well as their resultants. The SNR was estimated through Matlab 2022b software using the function r = snr(x) that returns the SNR in decibels relative to the carrier (dBc) (where r stands for SNR and x is the input signal, https://www.mathworks.com/help/signal/ref/snr.html, accessed on 24 December 2023). The function determines the SNR through a modified periodogram that uses a Kaiser window with $\beta = 38$ and has the same length as the input signal (the result excludes the power of the first six harmonics, including the fundamental). In breathing monitoring studies, an SNR as low as 10 dB has been realized as a limit above which the accuracy of results is consistently good [29]. To the best of our knowledge, breathing studies using inertial sensors do not report their SNR values; however, an SNR $\geq$ 14 dB is reported as producing very low estimation error concerning the breathing rate using a Doppler radar system [29] and typical of clinically acknowledged equipment such as the respiration-correlated computer tomography [30], while an average SNR at 29.6 dB is reported for acoustic breathing signal [31].

Table 2. Signal to noise ratio pre- and post-bandpass (0.1 Hz–0.5 Hz) signal filtering *.

Filtering Condition	Experimental Condition	Signal to Noise Ratio (dB)			
		Roll	Pitch	Yaw	Resultant
Pre-Filtering	EN	13	14	10	12
	ED	19	18	19	16
	HN	6	6	5	6
	HD	7	6	7	6
Post-Filtering	ED	93	110	101	96
	HN	47	41	38	36
	HD	44	39	44	37
	EN	79	81	69	64

* Matlab 2022b, https://www.mathworks.com/help/signal/ref/bandpass.html, accessed on 24 December 2023.

To ensure the removal of any remaining linear trend, signal filtering was followed by signal detrending using the function D = detrend(A) of MatLab R2022b software (https://www.mathworks.com/help/matlab/ref/detrend.html?s_tid=srchtitle_site_search_1_detrend, accessed on 10 August 2023) that removes the best straight-fit line from the data of the input signal (A) and returns the remaining data (D). Finally, an offset removal procedure was applied according to the equation y = x − offset, where x is the input signal and y is the output signal with the offset being the value of the initial data point of each respective input signal which was subtracted from each one of its successive data points. Sample pre- and post-filtering time series, as well as after offset removal, are illustrated in Figure 2 for ES and in Figure 3 for HS. Detrending did not affect the SNR of the time series signals or yield any significant difference concerning the variables under examination ($p > 0.05$). Thus, the detrended signals are not included in Figures 2 and 3.

All signals were plotted at each procession step for visual inspection. All three AD signals (roll, pitch, and yaw) presented peaks and dips allowing us to visualize the breathing cycles in both the normal and the diaphragmatic breathing conditions. Peaks indicate the end of the inspiration phase and dips at the end of the expiration one, thus the initiation of the subsequent breathing cycle. After locating the initiation of the first breathing cycle, a 30 s breathing duration was extracted for further analysis, concerning the roll, pitch, and yaw Euler angles and their resultants (in each one of the three trials per participant, for each body stance and breathing type experimental condition), as described below.

The initiation of the first breathing cycle was located through visual inspection and numerical verification of the first dip of the signal (end of first expiration). The abdomen wall rotation around the frontal horizontal axis traces the characteristic abdomen wall expansion and compression during inspiration and expiration, respectively, [20]. Thus, the roll AD signal was selected to visually locate the temporal position of the dip indicating the end of the first expiration, that is, the initial time point of the 30 s data used for further procession and analysis in all three AD signals. The visual location of the dip indicating the end of the first expiration was followed by verification that its numerical value was the lower one among the neighboring data. The three AD signals (roll, pitch, and yaw) were synchronized, and their visual inspection allowed us to assume a temporal concurrence among their dips and peaks; however, we did not extract, nor did we statistically test the dip and peak time differences among the roll, pitch, and yaw AD signals.

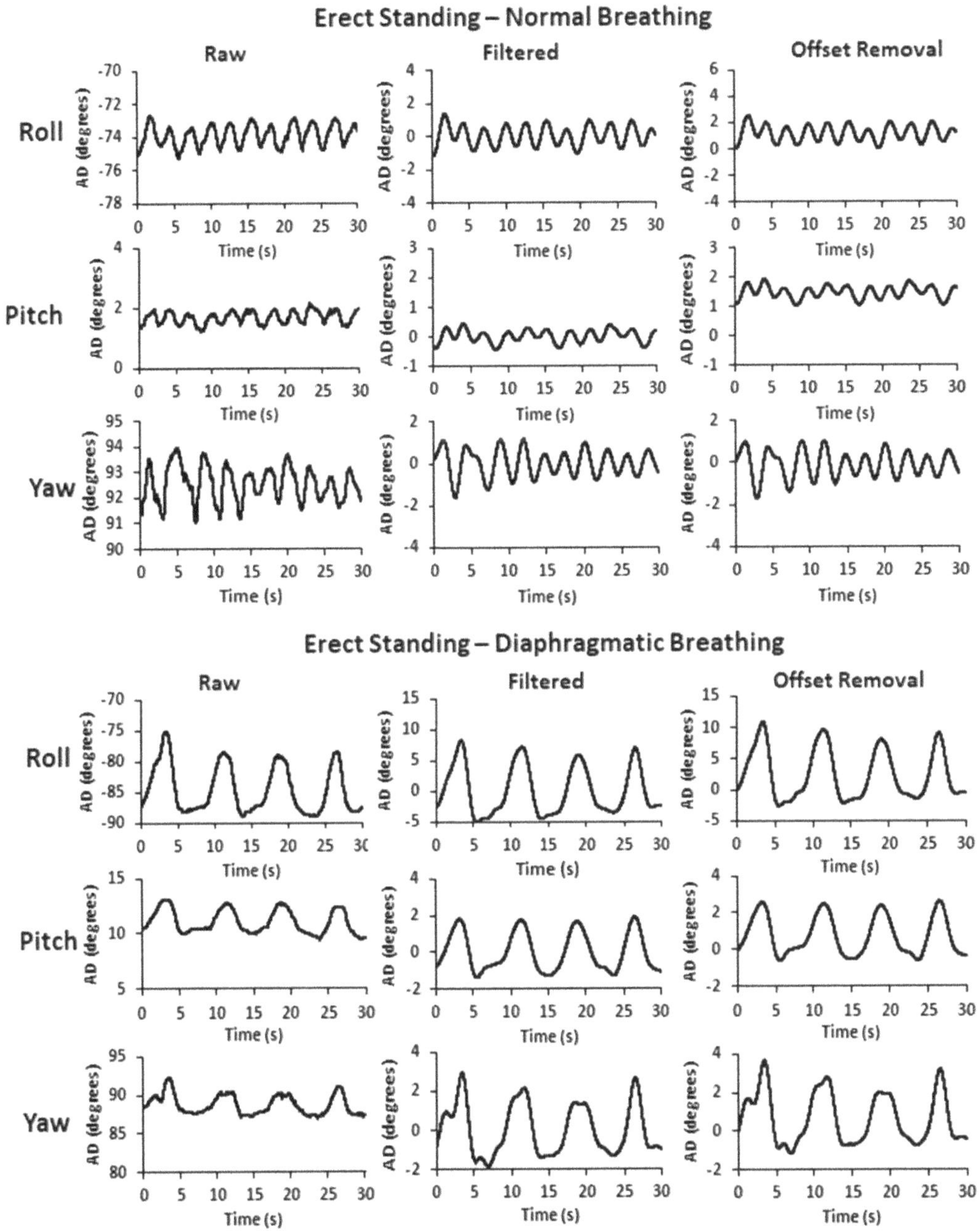

Figure 2. Sample pre- and post-filtering time series signals (raw and filtered, respectively), as well as after the offset removal, for the 3D angular displacement (roll, pitch, and yaw: forward, outward, and downward abdominal expansion, respectively) in erect standing during normal (**top**) and diaphragmatic (**bottom**) breathing.

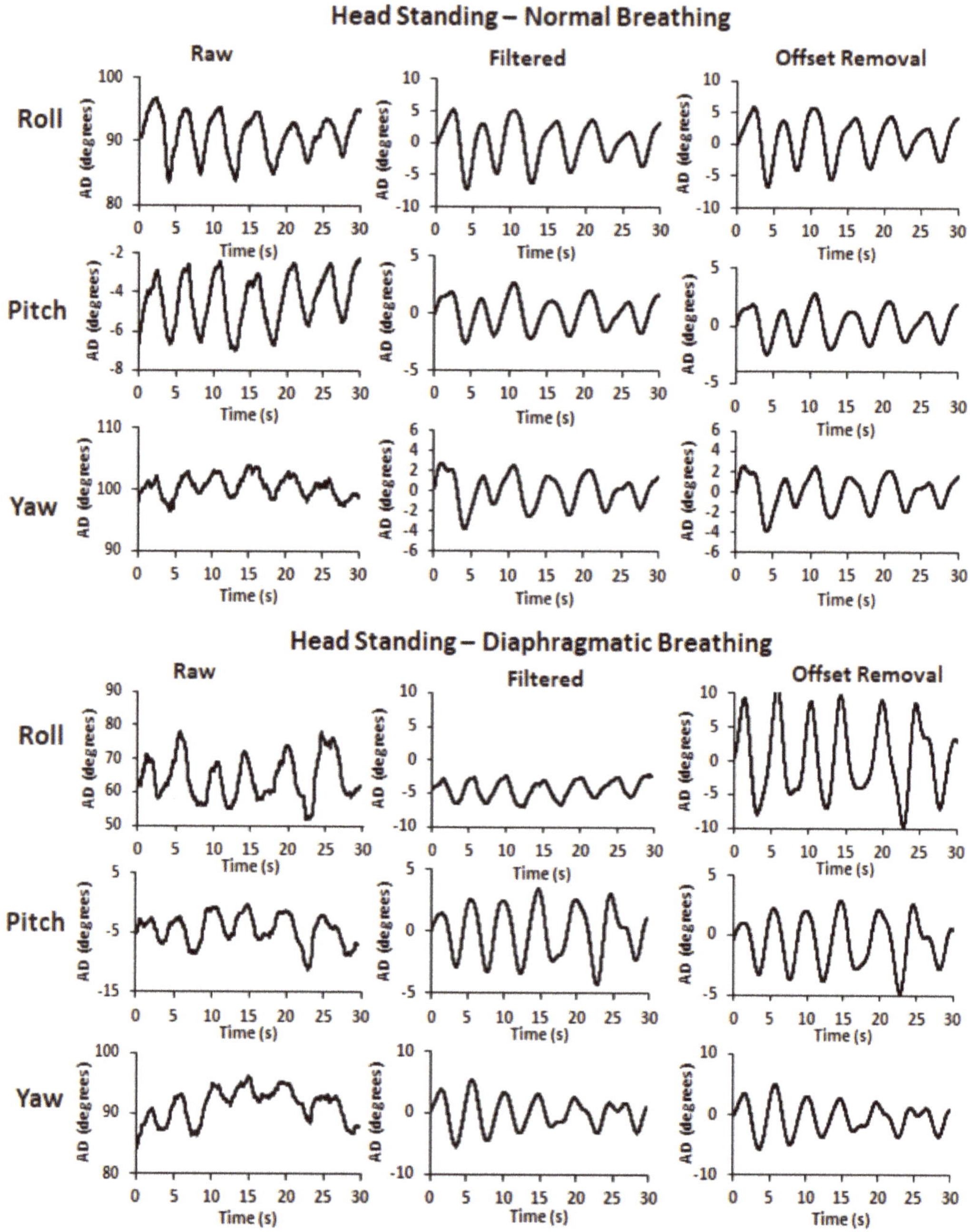

Figure 3. Sample pre- and post-filtering time series signals (raw and filtered, respectively), as well as after the offset removal, for the 3D angular displacement (roll, pitch, and yaw: forward, outward, and downward abdominal expansion, respectively) in head standing during normal (**top**) and diaphragmatic (**bottom**) breathing.

2.2.4. Estimation of Angular Displacement (AD) Variables

Magnitude of AD Path Length. The length of the AD path was calculated by summing the magnitude of the AD change (absolute value) at every time step for the data points comprising the AD time series signal. The AD path length was calculated by summing the magnitude of the AD distance change at every time step using Equation (1), where N is the number of data points in the AD time series (in this case, 3000 successive data points), θ is the angular position at the respective time step, and $|\cdot|$ denotes the absolute value. For each participant, the mean of his/her three trials per experimental condition consisted of the AD path length value inserted in statistical analysis.

$$\text{Magnitude of AD path length} = \sum_{n=2}^{N} |\theta[n] - \theta[n-1]| \quad (1)$$

Individual Variability of AD Path Length. Individual variability is a measure of natural behavior reliability [32,33]. Separately for each participant and the participant's three trials in each condition, the relative individual coefficient of variation (CVind) of the AD path length was calculated and expressed as a percentage (%CVind) according to Equation (2), where SD is the standard deviation and $\overline{x}$ is the average of the three trials' observed values.

$$\%CVind = \frac{SD}{\overline{x}} \times 100, \text{ where } CVind = \frac{SD}{\overline{x}} \quad (2)$$

Ratio of AD Path Length. For the AD path length, the diaphragmatic (D) to normal (N) breathing ratio was calculated, in ES (E-D/N) and HS (H-D/N), expressing the times that the AD of the abdomen wall was increased during the diaphragmatic relative to normal breathing.

2.2.5. Cross-Correlation, Relative Reliability, Standard Error of Measurement, and Minimal Detectable Change

Cross-Correlation of Time Series Signals. Separately for each participant, the signals extracted for further analysis were initially tested for cross-correlation between pairs of trials. Cross-correlation provides a measure of association between signals. When two time-series data sets are cross-correlated, a measure of temporal similarity or temporal reliability is achieved [34]. The cross-correlation coefficient (CC) was calculated between pairs of trials in each condition, that is, between trial 1 and trial 2 (CC12), between trial 1 and trial 3 (CC13), and between trial 2 and trial 3 (CC23). All CCs and their statistical significance were calculated using the MatLab Function [R,P] = corrcoef(A,B) (https://www.mathworks.com/help/matlab/ref/corrcoef.html, accessed on 10 August 2023), where R stands for the correlation coefficient, P for its significance level (default: P < 0.05), and A and B stand for the two discrete time series signals that are cross-correlated. The mathematical equations behind the corrcoef function concern the correlation coefficient matrix (R) (Equation (3)) and the *p*-value matrix (P) (Equations (4) and (5)). Concerning Equation (3), the cov(x,y) is the covariance between x and y and var(x) and var(y) are the variances of x and y, respectively.

$$Correlation\ Coefficient\ (R) = \frac{cov\ (x,y)}{\sqrt{\mathrm{var}(x)\cdot\mathrm{var}(y)}} \quad (3)$$

The *p*-value matrix was calculated using a t-statistic according to Equation (4), under the assumption that the true correlation coefficient is zero.

$$t = \frac{R\sqrt{n-2}}{\sqrt{1-R^2}} \quad (4)$$

The *p*-value was then obtained from the cumulative distribution function of the t-distribution with n − 2 degrees of freedom according to Equation (5); here, t(i,j) is the t-statistic for the correlation coefficient R(i,j) and P(i,j) is the *p*-value for testing the null hypothesis that the true correlation coefficient is zero.

$$P(i,j) = P(\text{abs}(t(i,j)) > \text{abs}(R(i,j))) \quad (5)$$

Relative Reliability–Intraclass Correlation Coefficient (ICC). The relative reliability of the AD path length among the three trials was tested using the Intraclass Correlation Coefficient (ICC). There are several forms of the ICC [35], but considering that the relative reliability that is in question concerns absolute agreement, the 2-way random effects absolute agreement method, also known as ICC (2,1), was the most appropriate for the present study [36], and the average measures output was used (SPSS v28.0). The ICC's upper and lower bounds of their 95% confidence interval were also extracted. Relative reliability was classified by Fleiss [37] (ICC > 0.75: excellent; ICC between 0.40 and 0.75: fair to good; ICC < 0.40: poor).

Standard Error of Measurement (SEM). A behavior change can be interpreted as a true change only if the observed difference between measurements is larger than the measurement error, the latter depending on the trials' absolute reliability [32,33] in addition to the natural behavior reliability (%IndCV). Absolute reliability was tested with the standard error of measurement (SEM) with a smaller SEM indicating a better absolute reliability. SEM accounts for the within-subject variability and assesses how precisely a test measures a subject's true value, has the same units as the measure of interest, and is not sensitive to the between-subject variability of the data. Thus, SEM indicates the expected variation in observed values that occurs owing to the measurement error (if reliability = 0, SEM will equal the standard deviation of the observed values; if test reliability = 1.00, SEM will be zero).

SEM was estimated as the square root of the mean square error term from ANOVA, as this estimation has the advantage of being independent of the specific ICC and allows more consistency in interpreting SEM values across different studies [38]. The relative SEM (expressed as a percentage, %SEM) was used for statistics and was defined as (SEM/$\overline{x}$ $\times$ 100, where $\overline{x}$ is the average of all participants' AD path length values. The %SEM allows a comparison of the expected variation in observed values between different conditions.

Minimal Detectable Change. The minimal detectable change (MDC) indicates the minimal amount of change that can be interpreted as a "real" change in the behavior of an individual, that is, a difference greater than the measurement error; a smaller MDC indicates a more sensitive measure [33,38]. MDC was calculated from the SEM based on a 90% confidence interval (MDC90) following the formula MDC90 = SEM $\times$ 1.65 $\times$ $\sqrt{2}$ [33]. Then, the relative MDC90 was estimated (%MDC90) as (MDC90/$\overline{x}$) $\times$ 100, where $\overline{x}$ is the average of all participants' observed values, separately for each combination of body stance and breathing type condition. The %MDC90 allowed the comparative evaluation of absolute reliability to infer if the observed differences could be interpreted as a "real" change in the behavior of the participants of the present study.

2.3. Statistical Analysis

A two-way repeated measures analysis of variance (ANOVA) (body stance X breathing type) was applied to test the interaction between body stance and breathing type concerning the magnitude, the %CVind, and the ratio of the AD path length. Due to non-significant interaction between body stance and breathing type across all three AD variables (Appendix A, Table A1, two separate one-way ANOVAs were applied, one to test the effect of breathing type per body stance and another one to test the effect of body stance per breathing type. The level of significance was set at $p \leq 0.05$ (SPSS version 28.0, IBM Statistics, Armonk, NY, USA).

3. Results

3.1. Angular Displacement Path Length

Across all axes of rotation, as well as their resultant, the AD path of the abdominal wall was significantly altered due to breathing type. Specifically, diaphragmatic breathing significantly increased the AD path of the abdominal wall ($p \leq 0.01$ for all) in both ES and HS (Figure 4). Similarly, the body stance altered the AD path of the abdominal wall, with a significant increase in HS for both normal and diaphragmatic breathing ($p \leq 0.01$ for all) (Figure 4). Detailed statistics of the repeated measures ANOVA, as well as the Cohen's d effect size, concerning the breathing type and the body stance main effect, are provided in Appendix A, Table A1.

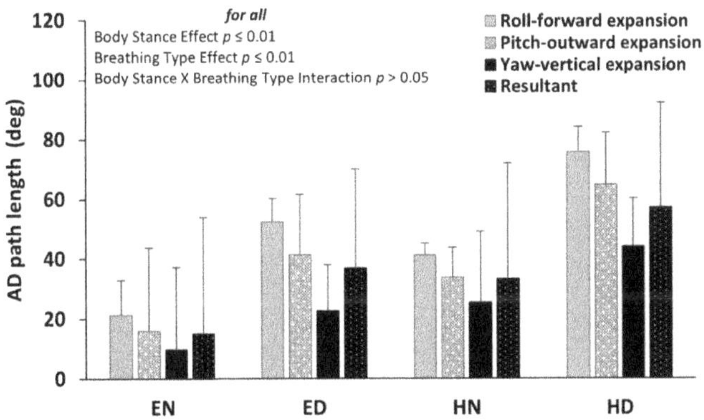

Figure 4. Mean (SD) of the angular displacement (AD) path length of the roll, pitch, and yaw Euler angles as well as their resultant for each condition of body stance and breathing type. The anatomical expansion of the abdomen wall during diaphragmatic inspiration is noted. The body stance effect and breathing type effect were both significant ($p \leq 0.01$ for all), with no significant interaction between them ($p > 0.05$). EN: Erect standing–Normal breathing, ED: Erect standing–Diaphragmatic breathing, HN: Head standing–Normal breathing, and HD: Head standing–Diaphragmatic breathing.

3.2. *Individual Variability of AD Path*

The relative individual coefficient of variation (%CVind) was not significantly altered due to either the body stance or due to breathing type ($p > 0.005$) (Figure 5). Detailed statistics of the repeated measures ANOVA, as well as the Cohen's d effect size, concerning the breathing type and the body stance main effect, are provided in Appendix A, Table A1.

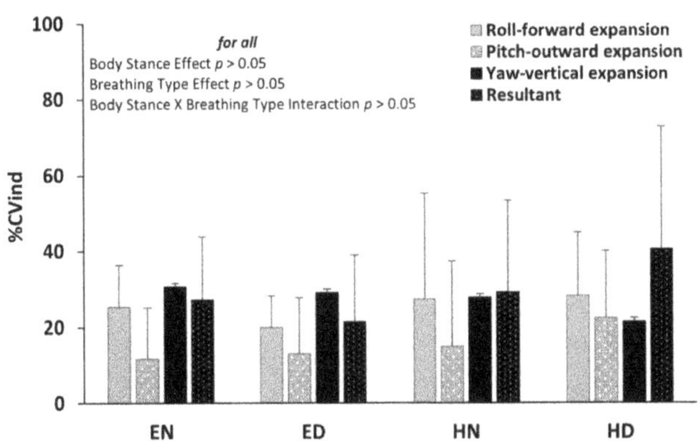

Figure 5. Mean (SD) of the individual coefficient of variation (%CVind) (bottom) of the roll, pitch, and yaw Euler angles, as well as their resultant, for each condition of body stance and breathing type. The anatomical expansion of the abdomen wall during diaphragmatic inspiration is noted. Neither the main effects of body stance and breathing type nor their interaction were significant ($p > 0.05$). EN: Erect standing–Normal breathing, ED: Erect standing–Diaphragmatic breathing, HN: Head standing–Normal breathing, and HD: Head standing–Diaphragmatic breathing.

3.3. Diaphragmatic to Normal Breathing Ratio of AD Path

Across all axes of rotation, as well as their resultants, the diaphragmatic to normal breathing ratio was significantly lower in HS ($p < 0.05$) indicating that it was tripled in ES and just doubled in HS (Figure 6). Detailed statistics of the repeated measures ANOVA, as well as the Cohen's d effect size, concerning the body stance main effect are provided in Appendix A, Table A2.

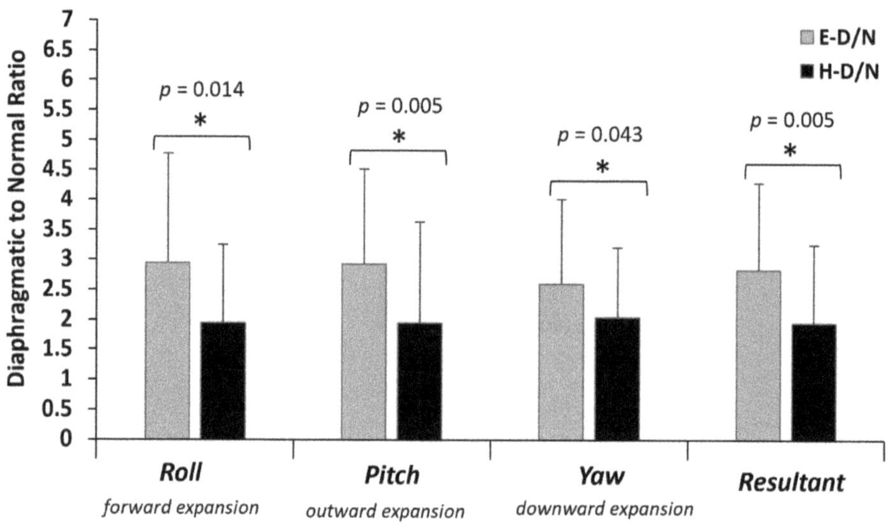

Figure 6. Diaphragmatic to normal breathing ratio of the angular displacement (AD) path length in erect standing (E-D/N: black bars) and head standing (H-D/N: grey bars). The anatomical expansion of the abdomen wall during diaphragmatic inspiration is noted. The ratio indicates the times that the AD path was increased in diaphragmatic compared to normal breathing. * significant difference between E-D/N and H-D/N at $p \leq 0.05$.

3.4. Cross-Correlation of AD Time Series Signals

The means of the pairwise cross-correlations for each body stance and breathing type are presented as boxplots in Figure 7. Overall, with a small number of non-consistent exceptions, all cross-correlations were significant (Figure 7).

3.5. Relative Reliability–Intraclass Correlation Coefficient (ICC) of AD Path

All ICCs (Figure 8-Right) were above 0.80 indicating excellent relative reliability of the AD path length in all three Euler angles as well as their resultants in all body stance X breathing type conditions.

3.6. Standard Error of Measurement and Minimal Detectable Change in AD Path

The %MDC90 was greater than the %SEM upper boundary limit (Figure 8-Left) indicating that the observed changes may be interpreted as "real" ones due to either body stance or due to breathing type.

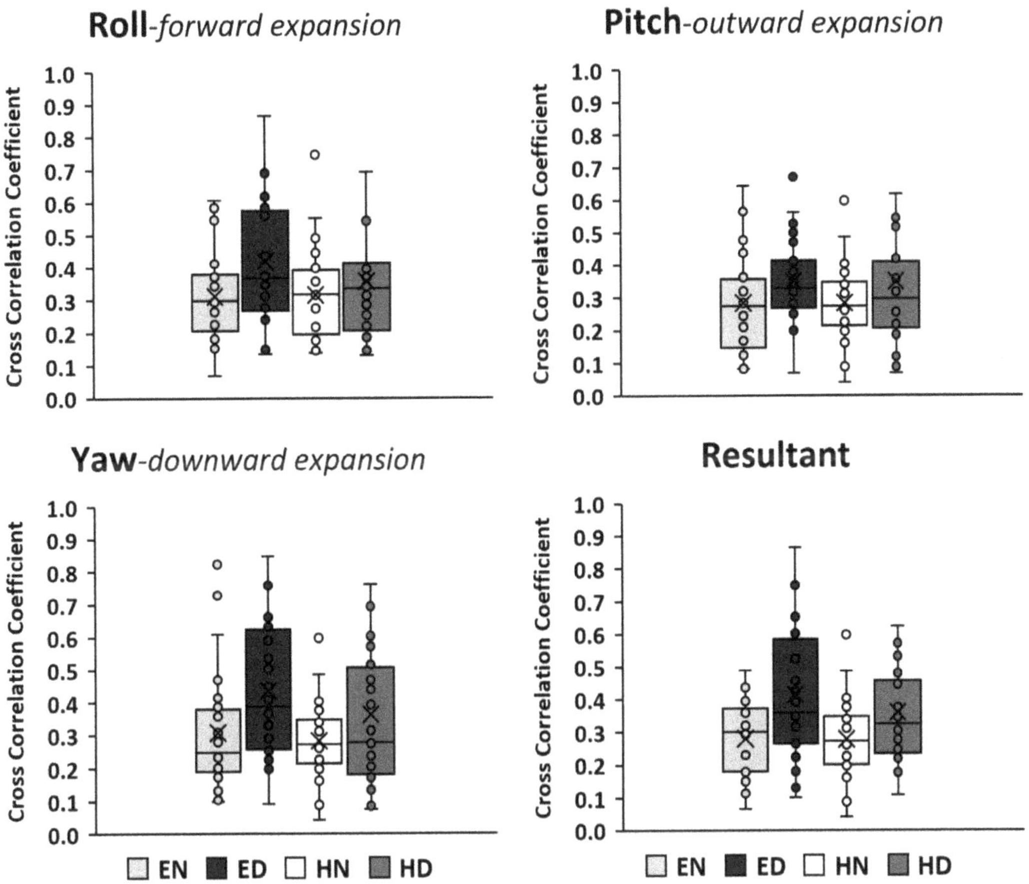

Figure 7. Boxplots of the mean coefficient of cross-correlation between pairs of trials in each condition. The box indicates the interquartile range of the values (IQR: 50% of the values lie within 0.6745 standard deviation). The horizontal line in the box indicates the median and the x symbol indicates the mean. The whiskers extend up from the top of the box to the largest data element that is less than or equal to 1.5 times the IQR and down from the bottom of the box to the smallest data element that is larger than 1.5 times the IQR. The filled circles indicate values, with those outside the whiskers considered as outliers. The anatomical expansion of the abdomen wall during diaphragmatic inspiration is noted. EN: Erect standing–Normal breathing, ED: Erect standing–Diaphragmatic breathing, HN: Head standing–Normal breathing, and HD: Head standing–Diaphragmatic breathing. Overall, with a very small number of exceptions, the cross-correlations were significant ($p \leq 0.05$).

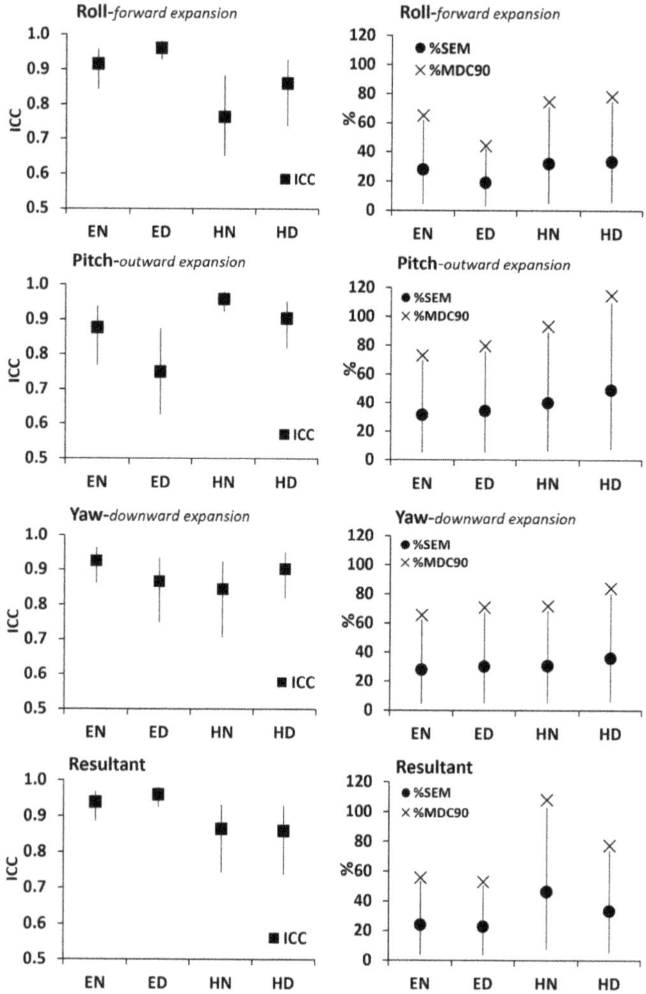

Figure 8. Right: Relative reliability of the AD path length (black squares indicate ICC with the vertical lines denoting the upper and lower bounds of its 95% confidence interval). **Left**: Absolute reliability of the AD path length (black circles indicate %SEM with the vertical lines denoting the upper and lower bounds of its 95% confidence interval) with the relative minimal detectable change (%MDC90) indicated by the × index. Results are presented for the roll, pitch, and yaw Euler angles as well as their resultants in EN: Erect standing–Normal breathing, ED: Erect standing–Diaphragmatic breathing, HN: Head standing–Normal breathing, and HD: Head standing–Diaphragmatic breathing.

4. Discussion

The purpose of this study was to apply inertial sensing to evaluate the effect of diaphragmatic versus normal breathing on the abdomen wall kinematics during head standing (HS). The initiative for the present study was that nowadays many people are encouraged or even challenged to practice HS as an atypical deviation from daily activity, (upsetting the evolved systemic adaptations for overcoming the effects of gravity), while concurrently practicing diaphragmatic breathing, the latter expected to further enhance the alleged health benefits of HS [7–9].

The present results, with no significant interaction between the body stance and the breathing type, verify a significant increase in the abdomen wall AD path length during diaphragmatic breathing compared to normal breathing, in both ES and HS ($p < 0.001$), across all three directions of abdominal expansion–compression. The results consolidate the visual criteria used to identify diaphragmatic breathing, that is, the greater displacement of the abdomen wall due to its inspiration-related expansion [10,11] followed by its active and intense, rather than the normal passive [12], flattening as the abdominal muscles contract against gravity to compress the viscera upward and inward.

Diaphragmatic breathing is known to affect the overall respiration mechanics [15] through diaphragm–related abdominal expansion. When breathing normally in the head-down position, the breathing mechanics are disturbed due to the reversed action of gravity on the ribcage (*inspiratory direction:* maximal gravity effect superseding the abdominal hydrostatic pressure, *expiratory direction:* reversed effect of gravity on the elastic recoil properties of the various contracting components of the respiratory muscles, less postural contraction of the respiratory muscles) [5].

While in HS, the lateral deviations in the thoracic spine may increase postural control difficulties as well as the spinal loads [39] adding to the disturbed normal breathing mechanics when gravity resists rather than aids the expiration phase. Foskolou et al. [14] report the postural stability differences between ES and HS in normal and diaphragmatic breathing. Overall, for normal and diaphragmatic breathing, HS appears to increase postural instability (anteroposterior and mediolateral) at about 8 to 9 times higher compared to ES [14]. Thus, in HS, the muscles have both a breathing and a postural role, such as the diaphragm and the abdominal muscles, and may prioritize their postural role to modulate the intra-thoracic and intra-abdominal pressure and therefore compromise the respiratory motion of the rib cage and the abdomen [11]. Indeed, the strong contraction of postural muscles in HS appears to play a major role in the greater air volume remaining in the lungs compared to hung head-down experiments [13].

One should also consider that the breathing waveform sensed by an accelerometer is a slow periodic variation (<1 Hz) with a weak amplitude, which is easily mixed with body movements (both signals are in the same frequency band [0, 1 Hz], so the person must stay still while taking measurements [28]. Although we may not exclude with absolute certainty the possibility of body movement interference, the signals of the present study have sustained an appropriate filtering process (passband at 0.1–0.5 cut-off frequencies) and agree with previously reported breathing waveforms [17–21,23].

Concerning the filter efficacy, it appears to have removed the data consisting of noise as shown by the improved SNR when comparing the pre- to post-filtering values. It must be noted though that even the pre-filtered SNR values appear to fall within the SNR range considered acceptable in breathing studies using a variety of breathing monitoring systems. Specifically, even 10 dB has been realized as an SNR limit above which the accuracy of results is consistently good [29]. Also, an SNR $\geq$ 14 dB is reported as producing a very low estimation error concerning the breathing rate using a Doppler radar system [29] and as being typical of clinically acknowledged equipment such as respiration-correlated computer tomography [30]. Also, an average SNR of 29.6 dB is reported for acoustic breathing signals [31]. To the best of our knowledge, there appears to be limited SNR information in breathing studies using inertial sensors. Phan et al. [28] using a sensor placed on the chest, suggest an adaptive rather than a fixed filter as potentially more effective regarding the SNR improvement; however, they do not report any SNR values. Overall, the average SNR of the present data, particularly those of the HS trials, was well within the commonly accepted SNR limits in previous breathing studies [29–31] and allowed us to assume a good efficacy of the applied filtering process. A potential consideration could be raised only for just a very small number of trials concerning solely the yaw AD signal during normal breathing in ES (sample signal provided in Figure 2 with an SNR at 10 dB).

One question that arises when inertial sensors are used to detect and evaluate breathing kinematics is the reliability of the results, which was also a concern of the present study. The reliability criteria themselves warrant careful computation and interpretation owing to the variety of computational models and the wide range of classification boundaries [35,37]. Furthermore, one should realize the different interpretations of the relative reliability (the degree to which individuals maintain their position in a sample over repeated measurements) versus absolute reliability indices (the degree to which repeated measurements vary for individuals) [37]. The present results are produced from AD time series of significant inter-trial temporal similarity, which together with the excellent relative reliability of the AD path length allow us to safely infer a high data reliability. The homogeneity of the testing conditions contributes to the reliability of the results [40,41]. Furthermore, despite the CV limitations as a reliability tool [42], the natural behavior reliability (expressed by %CVind [32,33]) was similar across experimental conditions which further enhances the overall data reliability.

Another issue that could affect the reliability of the results is the error of measurement, as any measured change reflects true change plus error. The error of measurement is the part of the obtained score that is unsystematic and random and due to chance, reflects the accumulated effects of all uncontrolled and unspecified influencing factors included in the obtained value [41]. When the error is great, wider confidence intervals apply to the observed difference, implying that the "true" change could be anywhere within the given range. The upper limit of this confidence interval helps to define another boundary of meaningful change. Theoretically, differences greater than this upper boundary limit would have less than a 5% chance of being changed due to chance (error) alone [36,41] and could therefore be confidently considered as a true difference due to the effect applied. Following this logic, the MDC also constitutes the lower boundary of a potentially meaningful change.

The acceptable error of measurement does not depend on the absolute SEM value, but should rather be considered with regards to the measured parameter, as well as with regards to the MDC of the measure under evaluation. To the best of our knowledge, there do not appear standardized specific criteria concerning the %SEM classification. As emphasized by Atkinson and Nevill [42], the higher the %SEM, the lower the absolute reliability and the lower the precision of the obtained results. A %SEM of 9–14% was classified by Jaworski et al. [43] as very good reliability (single-legged standing, Gyko inertial sensor). Pooranawatthanakul and Siriphorn [44] reported %SEM at about 16% in common ES balance tests (mobile phone accelerometer). The %SEM in the present study (ranging from 18.9 to 49.2% with an average, across-body stance and breathing type conditions) together with the about 5% higher %MDC90 than the upper boundary limit of the %SEM, allows us to safely assume that the observed differences reflect reliable as well as "real" changes.

One could argue for the use of inertial sensors rather than the gold standards for testing breathing function (spirometry and body plethysmography, Beck et al. [17]) due to the problems associated with the use of inertial sensors to detect and evaluate the breathing kinematics [20,24]. However, both spirometry and body plethysmography are rather uncomfortable as the person must breathe through a mouthpiece or a face mask with a sealed nose [17], an issue that initiated studies dated back to the middle of the 20th century to measure respiratory parameters by focusing on movements of the thorax [45]. Nowadays, inertial sensing of the breathing kinematics has been validated against plethysmography [17,19], spirometry [20], pneumotachography [21], and nasal pressure [16], whereas the body placement or motion artifact problems [20,24] appear to be effectively handled through elaborated signal-filtering techniques [17,23,25].

It is important to note that the accelerometer can introduce some systematic errors, even under identical experimental conditions [27]. However, we tried to eliminate the influence of external factors by maintaining the same experimental conditions for all subjects (conducting the data collection at fixed times of the day with some tolerance, monitoring air temperature and humidity to keep similar ambient conditions, and ensuring the absence of environmental sound or voice pollutants—a possible reason for a variety of oscillations).

One could also argue that the sensor body placement (abdominal rather than chest wall) as the body location is considered to affect the results. During normal breathing in ES, both the chest [16,17,19,20,22,23] and the abdomen wall [16,18–21,23] placements appear to effectively detect kinematic changes (inclination angle and rotation). However, the studies favoring the sensor placement on the chest wall [16,17,19,20,22,23] examine normal rather than diaphragmatic breathing, which justifies their choice for chest wall placement. Specifically, Karacocuk et al. [20] mention that in their experimental condition, abdominal breathing dominated instead of thoracic breathing, and they could not detect the abdominal breathing kinematic changes because the sensor was placed on the chest.

In the present study, the sensor was placed on the abdomen wall at the umbilicus level rather than closer to the rib cage applied in other studies [18,23]. Siqueira et al. [21] used 10 sensor positions (spread vertically as well as horizontally on the frontal body surface) to investigate the best locations that lead to good quality signals (3 sensors in the middle of the chest using the nipples as a reference, 3 sensors on the costal arch, and 3 sensors on the abdomen using the umbilicus as a reference, including the body position of the present study, and one sensor on the abdomen at the same height as the ones placed on the lower ends of the costal arch). Taking into consideration that Siqueira et al. [21] studied normal breathing in conventional body stances (standing, sitting, and lying), they reported as best locations those on the right body side (most likely due to the greater lung size on the right) and the apex of the costal arch. Nevertheless, as shown by Siqueira et al. [21], the sensor placement used in the present study is effective in detecting the kinematic changes in the body surface due to breathing, with improved efficacy in the lying than in the standing or sitting body stance. The rationale for placing the sensor near the umbilicus in the present study was that this is the anatomical level where the greatest abdomen expansion is expected during diaphragmatic inspiration both vertically and horizontally [20]. In addition, Karacocuk et al. [20] attributed their inability to detect the abdominal breathing contribution to their experimental results because their sensor was placed on the chest rather than the abdomen, the latter also avoiding the heart rate [28] influence on the signal. As both men and women were participants in the study (indeed, 72% of them were women), the abdomen rather than the chest location also allowed the avoidance of the potential breast [21] influence in the kinematic changes due to breathing. Potential biases emanating from the dominance of women participants may not be excluded as the breathing function is associated with differences due to gender (most likely due to their smaller size than men and possibly as pregnancy propaedeutics) [46]. However, considering all reliability indices (%CVind, ICC, %SEM, and %MDC90), we feel safe to state that neither the sensor placement nor the participants' gender appears to have undermined the accuracy or the precision of the results.

5. Conclusions

In conclusion, the present study examined the effect of diaphragmatic breathing in the HS position in comparison to both normal breathing as well as ES. The results indicated a significant increase in the abdomen wall AD during diaphragmatic breathing compared to normal breathing in HS, whereas no significant interaction was observed between body stance and breathing type. This finding provides experimental and quantitative verifications of the visual criteria used to identify diaphragmatic breathing, i.e., the abdomen expansion and depression during the diaphragmatic inspiration and expiration phase, respectively. The increase in the AD path length in diaphragmatic breathing was significantly lower in HS than in ES most possibly due to the breathing mechanics alterations when the body is inverted head-down (mainly due to gravity-related differences associated with the interplay between the breathing and postural role of the diaphragm as the primary inspiratory muscle and the muscles of the abdomen wall as the primary expiratory muscles.

Across all body stance and breathing type conditions, the results appear to be highly reliable as shown by the temporal similarity of the AD time series obtained through the cross-correlation analysis, the natural behavior reliability examined through the %CVInd, the relative reliability measures examined through the ICC, and the absolute reliability examined through the %SEM. The small number of non-significant cross-correlations may not be explained as these did not yield a consistent pattern across pairs of trials or for AD direction, body stance, breathing type, or participant. Finally, the %MDC90 being consistently higher (about 5%) than the upper boundary limit of the %SEM allows us to safely conclude the AD differences as "real" changes due to diaphragmatic breathing, in both ES and HS. Our study contributes to research-based evidence concerning the concurrent practice of diaphragmatic breathing during HS, as, nowadays, many people are encouraged or even challenged to engage in this practice to further enhance the alleged benefits of HS.

Author Contributions: Conceptualization, E.R. and A.F.; methodology, E.R. and A.F.; formal analysis, E.R. and A.F.; investigation, E.R. and A.F.; data curation, E.R. and A.E.; writing—original draft preparation, E.R., A.E. and A.F.; writing—review and editing, E.R., A.E. and K.B.; visualization, A.E. and E.R.; supervision, E.R. and K.B.; project administration, K.B. and E.R. All authors have read and agreed to the published version of the manuscript.

Funding: This research received no external funding.

Institutional Review Board Statement: The study was conducted in accordance with the Declaration of Helsinki and approved by the Bioethics Committee of the School of Physical Education and Sport Science, National and Kapodistrian University of Athens, Greece (no. 1016/14/02/2018).

Informed Consent Statement: Informed consent was obtained from all subjects involved in the study.

Data Availability Statement: The data are not publicly available due to ethical restrictions.

Acknowledgments: We thank all participants for the contribution in this work.

Conflicts of Interest: The authors declare no conflict of interest.

Appendix A

Table A1. Main effect (body stance and breathing type) and interaction (body stance and breathing type) statistics for the two-way repeated measures ANOVA concerning the angular displacement path length (AD path length) and the individual variability of the AD path (%CVind) for each one of the three Euler angles (roll, pitch, and yaw) as well as their resultants.

		F	Significance p Value	Partial Eta Squared	Cohen's d Effect Size	Observed Power
			Body Stance Main Effect			
AD path length	Roll	21.60	0.000 *	0.444	1.79	0.99
	Pich	23.82	0.000 *	0.469	1.88	1.00
	Yaw	14.35	0.001 *	0.347	1.46	0.95
	Resultant	22.52	0.000 *	0.455	1.83	1.00
%CVind	Roll	3.75	0.063 ^{ns}	0.122	0.75	0.46
	Pich	2.88	0.101 ^{ns}	0.097	0.65	0.37
	Yaw	1.48	0.234 ^{ns}	0.052	0.47	0.22
	Resultant	4.76	0.038 ^{ns}	0.150	0.84	0.56
			Breathing Type Main Effect			
AD path length	Roll	39.70	0.000 *	0.595	2.43	1.00
	Pich	64.23	0.000 *	0.704	3.09	1.00
	Yaw	15.70	0.000 *	0.368	1.53	0.97
	Resultant	35.75	0.000 *	0.570	2.30	1.00
%CVind	Roll	0.49	0.489 ^{ns}	0.018	0.27	0.10
	Pich	1.63	0.212 ^{ns}	0.057	0.49	0.23
	Yaw	0.82	0.374 ^{ns}	0.029	0.35	0.14
	Resultant	0.33	0.571 ^{ns}	0.012	0.22	0.09

Table A1. *Cont.*

		F	Significance p Value	Partial Eta Squared	Cohen's d Effect Size	Observed Power
			Body Stance * Breathing Type Interaction			
AD path length	Roll	0.10	0.754 ns	0.004	0.12	0.06
	Pich	0.85	0.364 ns	0.031	0.36	0.15
	Yaw	0.72	0.404 ns	0.026	0.33	0.13
	Resultant	0.12	0.734 ns	0.004	0.13	0.06
%CVind	Roll	1.22	0.279 ns	0.043	0.43	0.19
	Pich	0.51	0.483 ns	0.018	0.27	0.11
	Yaw	0.22	0.640 ns	0.008	0.18	0.07
	Resultant	5.74	0.024 ns	0.175	0.92	0.64

* Significant p value at $p \leq 0.05$, ns non-significant p value at $p \leq 0.05$. Commonly used interpretation of Cohen's d effect size [47]: small (d = 0.2), medium (d = 0.5), and large (d = 0.8) based on benchmarks suggested by Cohen [48]. Another way to interpret Cohen's d is the standard deviations (SDs) that the two groups differ between them, i.e., d = 0.05, d = 1, and d= 2, indicate, respectively, that the two group means differ by 0.5 SD, 1 SD, and 2SD Cohen [48].

Table A2. Statistics for the stance effect of the one-way repeated measures ANOVA concerning the diaphragmatic to normal breathing ratio of the AD path for each one of the three Euler angles (roll, pitch, and yaw) as well as their resultants.

	F	Significance p Value	Partial Eta Squared	Cohen's d Effect Size	Observed Power
Roll	6.83	0.014 *	0.202	1.00	0.71
Pitch	9.18	0.005 *	0.254	1.17	0.83
Yaw	4.50	0.043 *	0.143	0.82	0.53
Resultant	9.50	0.005 *	0.259	1.18	0.84

* Significant p value at $p \leq 0.05$, Commonly used interpretation of Cohen's d effect size [47]: small (d = 0.2), medium (d = 0.5), and large (d = 0.8) based on benchmarks suggested by Cohen [48]. Another way to interpret Cohen's d is the of standard deviations (SDs) that the two groups differ between them, i.e., d = 0.05, d = 1, d = 2, indicate, respectively, that the two group means differ by 0.5 SD, 1 SD, 2SD Cohen [48].

References

1. Liu, H.; Xu, Q.; Xiang, X.; Liu, D.; Si, S.; Wang, L.; Lv, Y.; Liao, Y.; Yang, H. Case report: Passive handstand promotes cerebrovascular elasticity training and helps delay the signs of aging: A 40-year follow-up investigation. *Front. Med.* **2022**, *9*, 752076. [CrossRef] [PubMed]
2. Manjunath, N.K.; Telles, S. Effects of sirsasana (headstand) practice on autonomic and respiratory variables. *Indian J. Physiol. Pharmacol.* **2003**, *47*, 34–42. Available online: https://ijpp.com/IJPP%20archives/2003_47_1/vol47_no1_orgn_artcl_2.htm (accessed on 10 August 2023). [PubMed]
3. Rao, S. Metabolic cost of head-stand posture. *J. Appl. Physiol.* **1962**, *17*, 117–118. [CrossRef] [PubMed]
4. Rao, S. Cardiovascular responses to head-stand posture. *J. Appl. Physiol.* **1963**, *18*, 987–990. [CrossRef] [PubMed]
5. Rao, S. Respiratory responses to headstand posture. *J. Appl. Physiol.* **1968**, *24*, 697–699. [CrossRef] [PubMed]
6. LeMarr, J.D.; Golding, A.L.; Kevin, D.; Crehan, K.D. Cardiorespiratory responses to inversion. *Phys. Sportsmed.* **1983**, *11*, 51–57. [CrossRef] [PubMed]
7. Iyengar, B.K. *Light on Yoga*; Schocken Books: New York, NY, USA, 1966.
8. Ma, X.; Yue, Z.-Q.; Gong, Z.-Q.; Zhang, H.; Duan, N.-Y.; Shi, Y.-T.; Wei, G.-X.; Li, Y.-F. The effect of diaphragmatic breathing on attention, negative affect and stress in healthy adults. *Front. Psychol.* **2017**, *8*, 234806. [CrossRef] [PubMed]
9. Martarelli, D.; Cocchioni, M.; Scuri, S.; Pompei, P. Diaphragmatic breathing reduces exercise-induced oxidative stress. *Evid. Based. Complement. Altern. Med.* **2011**, *2011*, 932430. [CrossRef]
10. De Troyer, A.; Estenne, M. Functional anatomy of the respiratory muscles. *Clin. Chest Med.* **1988**, *9*, 175–193. [CrossRef]
11. Hodges, P.W.; Gandevia, S.C. Activation of the human diaphragm during a repetitive postural task. *J. Physiol.* **2000**, *522*, 165–175. [CrossRef]
12. Fogarty, M.J.; Mantilla, C.B.; Sieck, G.C. Breathing: Motor Control of Diaphragm Muscle. *Physiol* **2018**, *33*, 113–126. [CrossRef] [PubMed]
13. Agostoni, E.; Hyatt, R.E. Static behavior of the respiratory system. In *Comprehensive Physiology, Supplement 12 Handbook of Physiology, the Respiratory System, Mechanics of Breathing*; Maclem, P.T., Mead, J., Eds.; American Physiological Society: Bethesda, MD, USA, 1986; pp. 113–130. [CrossRef]
14. Foskolou, A.; Emmanouil, A.; Boudolos, K.; Rousanoglou, E. Abdominal breathing effect on postural stability and the respiratory muscles' activation during body stances used in fitness modalities. *Biomechanics* **2022**, *2*, 478–493. [CrossRef]
15. Perry, S.F.; Similowski, T.; Klein, W.; Codd, J.R. The evolutionary origin of the mammalian diaphragm. *Respir. Physiol. Neurobiol.* **2010**, *171*, 1–16. [CrossRef] [PubMed]

16. Bates, A.; Ling, M.; Mann, J.; Arvind, D.K. Respiratory rate and flow waveform estimation from tri-axial accelerometer data. In Proceedings of the 2010 International Conference on Body Sensor Networks, Singapore, 7–9 June 2010; IEEE Computer Society: Singapore, 2010; pp. 144–150. [CrossRef]
17. Beck, S.; Laufer, B.; Krueger-Ziolek, S.; Moeller, K. Measurement of respiratory rate with inertial measurement units. *Curr. Dir. Biomed. Eng.* **2020**, *6*, 237–240. [CrossRef]
18. Erfianto, B.; Rizal, A. IMU-based respiratory signal processing using cascade complementary filter. *J. Sens.* **2022**, *2022*, 798715 [CrossRef]
19. Hung, P.D.; Bonnet, S.; Guillemaud, R.; Castelli, E.; Yen, P.T.N. Estimation of respiratory waveform using an accelerometer. In Proceedings of the 5th IEEE International Symposium on Biomedical Imaging: From Nano to Macro, Paris, France, 14–17 May 2008; pp. 1493–1496. Available online: https://ieeexplore.ieee.org/document/4541291 (accessed on 10 August 2023).
20. Karacocuk, G.; Höflinger, F.; Zhang, R.; Reindl, L.M.; Laufer, B.; Möller, K.; Röell, M.; Zdzieblik, D. Inertial sensor-based respiration analysis. *IEEE Trans. Instrum. Meas.* **2019**, *68*, 4268–4275. [CrossRef]
21. Siqueira, A.; Spirandeli, F.; Moraes, R.; Zarzoso, V. Respiratory waveform estimation from multiple accelerometers: An optimal sensor number and placement analysis. *IEEE J. Biomed. Health Inform.* **2018**, *23*, 1507–1515. [CrossRef]
22. Gaidhani, A.; Moon, K.S.; Ozturk, Y.; Lee, S.Q.; Youm, W. Extraction and analysis of respiratory motion using wearable inertial sensor system during trunk motion. *Sensors* **2017**, *17*, 2932. [CrossRef]
23. Hughes, S.; Liu, H.; Zheng, D. Influences of sensor placement site and subject posture on measurement of respiratory frequency using triaxial accelerometers. *Front Physiol.* **2020**, *11*, 823. [CrossRef]
24. Chu, M.; Nguyen, T.; Pandey, V.; Zhou, Y.; Pham, H.N.; Bar-Yoseph, R.; Radom-Aizik, S.; Jain, R.; Cooper, D.M.; Khine, M. Respiration rate and volume measurements using wearable strain sensors. *NPJ Digit. Med.* **2019**, *2*, 8. [CrossRef]
25. Presti, D.L.; Massaroni, C.; Caponero, M.; Formica, D.; Schena, E. Cardiorespiratory monitoring using a mechanical and an optical system. In Proceedings of the IEEE International Symposium on Medical Measurements and Applications (MeMeA), Lausanne, Switzerland, 23–25 June 2021; pp. 1–6. [CrossRef]
26. Lee, H.-Y.; Cheon, S.-H.; Yong, M.-S. Effect of diaphragm breathing exercise applied on the basis of overload principle. *J. Phys. Ther. Sci.* **2017**, *29*, 1054–1056. [CrossRef] [PubMed]
27. Boiko, A.; Gaiduk, M.; Scherz, W.D.; Gentili, A.; Conti, M.; Orcioni, S.; Martínez Madrid, N.; Seepold, R. Monitoring of cardiorespiratory parameters during sleep using a special holder for the accelerometer sensor. *Sensors* **2023**, *23*, 5351. [CrossRef] [PubMed]
28. Phan, D.H.; Bonnet, S.; Guillemaud, R.; Castelli, E.; Pham Thi, N.Y. Estimation of respiratory waveform and heart rate using an accelerometer. In Proceedings of the 30th Annual International Conference of the IEEE Engineering in Medicine and Biology Society, Vancouver, BC, Canada, 20–25 August 2008; pp. 4916–4919. [CrossRef]
29. Droitcour, A.D.; Boric-Lubecke, O.; Kovacs, G.T.A. Signal-to-noise ratio in doppler radar system for heart and respiratory rate measurements. *IEEE Trans. Microw. Theory Tech.* **2009**, *57*, 2498–2507. [CrossRef]
30. Heinz, C.; Reiner, M.; Belka, C.; Walter, F.; Söhn, M. Technical evaluation of different respiratory monitoring systems used for 4D CT acquisition under free breathing. *J. Appl. Clin. Med. Phys.* **2015**, *16*, 334–349. [CrossRef] [PubMed]
31. Corbishley, P.; Rodriguez-Villegas, E. Breathing detection: Towards a miniaturized, wearable, battery-operated monitoring system. *IEEE. Trans. Biomed. Eng.* **2008**, *55*, 196–204. [CrossRef] [PubMed]
32. Beaton, D.E.; Bombardier, C.; Katz, J.N.; Wright, J.G. A taxonomy for responsiveness. *J. Clin. Epidemiol.* **2001**, *54*, 1204–1217 [CrossRef] [PubMed]
33. Dontje, M.L.; Dall, P.M.; Skelton, D.A.; Gill, J.M.R.; Chastin, S.F.M.; Seniors USP Team. Reliability, minimal detectable change and responsiveness to change: Indicators to select the best method to measure sedentary behaviour in older adults in different study designs. *PLoS ONE* **2018**, *13*, e0195424. [CrossRef] [PubMed]
34. Derrick, T.R.; Thomas, J.M. Time-series analysis: The cross-correlation function. In *Innovative Analyses of Human Movement*, 1st ed. Stergiou, N., Ed.; Human Kinetics Publishers: Champaign, IL, USA, 2004; pp. 189–205.
35. McGraw, K.O.; Wong, S.P. Forming inferences about some intraclass correlation coefficients. *Psychol. Methods* **1996**, *1*, 30–46 [CrossRef]
36. Pinsault, N.; Vuillerme, N. Test-retest reliability of centre of foot pressure measures to assess postural control during unperturbed stance. *Med. Eng. Phys.* **2009**, *31*, 276–286. [CrossRef]
37. Fleiss, J.L. Reliability of measurement. In *The Design and Analysis of Clinical Experiments*; Fleiss, J.L., Ed.; Wiley: New York, NY USA, 1986; pp. 1–32.
38. Weir, J.P. Quantifying test-retest reliability using the intraclass correlation coefficient and the SEM. *J. Strength Cond. Res.* **2005**, *19* 231–240. [CrossRef]
39. Campos, M.H.; Giraldi, N.M.; Gentil, P.; de Lira, C.A.; Vieira, C.A.; de Paula, M.C. The geometric curvature of the spine during the sirshasana, the yoga's headstand. *J. Sports Sci.* **2017**, *35*, 1134–1141. [CrossRef]
40. Hopkins, W.G. Measures of reliability in sports medicine and science. *Sports Med.* **2000**, *30*, 1–15. [CrossRef]
41. Furlan, L.; Sterr, A. The applicability of standard error of measurement and minimal detectable change to motor learning research-A behavioral study. *Front. Hum. Neurosci.* **2018**, *12*, 95. [CrossRef] [PubMed]
42. Atkinson, G.; Nevill, A.M. Statistical methods for assessing measurement error (reliability) in variables relevant to sports medicine *Sports Med.* **1998**, *26*, 217–238. [CrossRef]

43. Jaworski, J.; Ambroży, T.; Lech, G.; Spieszny, M.; Bujas, P.; Żak, M.; ChwaŁa, W. Absolute and relative reliability of several measures of static postural stability calculated using a GYKO inertial sensor system. *Acta Bioeng. Biomech.* **2020**, *22*, 94–99. [CrossRef] [PubMed]
44. Pooranawatthanakul, K.; Siriphorn, A. Comparisons of the validity and reliability of two smartphone placements for balance assessment using an accelerometer based application. *Eur. J. Physiother.* **2020**, *22*, 236–242. [CrossRef]
45. Konno, K.; Mead, J. Measurement of the separate volume changes of rib cage and abdomen during breathing. *J. Appl. Physiol.* **1967**, *22*, 407–422. [CrossRef] [PubMed]
46. LoMauro, A.; Aliverti, A. Sex differences in respiratory function. *Breathe* **2018**, *14*, 131–140. [CrossRef]
47. Lakens, D. Calculating and reporting effect sizes to facilitate cumulative science: A practical primer for t-tests and ANOVAs. *Front Psychol.* **2013**, *4*, 863. [CrossRef]
48. Cohen, J. *Statistical Power Analysis for the Behavioral Sciences*, 2nd ed.; Lawrence Erlbaum Associates Inc.: Mahwah, NJ, USA, 1988; pp. 24–27. [CrossRef]

biomechanics

MDPI

Article

An Automated Approach to Instrumenting the Up-on-the-Toes Test(s)

Sarah Aruje Zahid [1], Yunus Celik [2], Alan Godfrey [2] and John G. Buckley [1,*]

1 Department of Biomedical and Electronics Engineering, Faculty of Engineering and Informatics, University of Bradford, Bradford BD7 1DP, UK; s.a.zahid@bradford.ac.uk
2 Department of Computer and Information Sciences, University of Northumbria, Newcastle upon Tyne NE1 8ST, UK; yunus.celik@northumbria.ac.uk (Y.C.); alan.godfrey@northumbria.ac.uk (A.G.)
* Correspondence: j.buckley@bradford.ac.uk

Abstract: Normal ankle function provides a key contribution to everyday activities, particularly step/stair ascent and descent, where many falls occur. The rising to up-on-the-toes (UTT) 30 second test (UTT-30) is used in the clinical assessment of ankle muscle strength/function and endurance and is typically assessed by an observer counting the UTT movement completed. The aims of this study are: (i) to determine whether inertial measurement units (IMUs) provide valid assessment of the UTT-30 by comparing IMU-derived metrics with those from a force-platform (FP), and (ii) to describe how IMUs can be used to provide valid assessment of the movement dynamics/stability when performing a single UTT movement that is held for 5 s (UTT-stand). Twenty adults (26.2 ± 7.7 years) performed a UTT-30 and a UTT-stand on a force-platform with IMUs attached to each foot and the lumbar spine. We evaluate the agreement/association between IMU measures and measures determined from the FP. For UTT-30, IMU analysis of peaks in plantarflexion velocity and in FP's centre of pressure (CoP) velocity was used to identify each repeated UTT movement and provided an objective means to discount any UTT movements that were not completed 'fully'. UTT movements that were deemed to have not been completed 'fully' were those that yielded peak plantarflexion and CoP velocity values during the period of rising to up-on-the-toes that were below 1 SD of each participant's mean peak rising velocity across their repeated UTT. The number of UTT movements detected by the IMU approach (23.5) agreed with the number determined by the FP (23.6), and each approach determined the same number of 'fully' completed movements (IMU, 19.9; FP, 19.7). For UTT-stand, IMU-derived movement dynamics/postural stability were moderately-to-strongly correlated with measures derived from the FP. Our findings highlight that the use of IMUs can provide valid assessment of UTT test(s).

Keywords: inertial measurement unit; feature extraction; ankle function; dynamic postural control; up-on-the-toes

1. Introduction

The functions of the foot and ankle are important to everyday locomotion/functional ability [1–3], including step/stair descent and ascent [4,5]. For example, the ankle provides a key contribution to maintaining dynamic balance during the single-limb support phase of stair descent when the contralateral limb is being lowered to contact the next (lower) step [4]. The rising to up-on-the-toes (UTT) 30 s test (UTT-30, also referred to as the calf-raise or heel-rise test) is used in the clinical assessment of ankle muscle strength, performance, and endurance [6] and ultimately the properties of the calf muscle–tendon unit [7]. The test has been deployed in the assessment of impairment, injury, and treatment outcomes of the lower limb [7]. The test assesses a person's ability to repetitively rise-UTT as many times as possible in 30 s and is typically timed using a stopwatch and assessed by an observer.

To encourage the patient to complete each repeated rising-UTT repetition over a 'full' (acceptable) range of motion (ROM), various monitoring devices have been developed. For example, using an electrogoniometer to measure ankle motion, 50% of each participant's maximum ankle ROM as determined prior to testing was used as the target ROM to indicate acceptable UTT movements [8,9]. Another study deployed a height-adjustable elastic band device placed beneath one heel, the height of which was adjusted to the height of the heel during a single UTT movement completed over a maximum comfortable height. Failure of the heel to then clear the band on two consecutive movements ended the UTT-30 test [10]. Elsewhere, monitoring of UTT movements has been achieved by placing a foam-board above the participant's head, the height of which was adjusted to the height attained when completing a single, controlled UTT movement that was as high as comfortably possible. Participants were asked to touch the board with their head on each UTT movement of the UTT-30 [6]. Although the use of such devices may encourage patients to complete the repeated rising-UTT movements over a 'full' (acceptable) ROM, they typically require an observer to keep track of the monitoring device to determine which repetitions are completed over an acceptable range and to count how many are completed. This may explain why none of these devices have become widely used.

In addition, whilst in past studies the inter-rater reliability for determining/observing the number of UTT repetitions completed in a UTT-30 has been shown to be high (ICC = 0.93–0.96) [6], a poorer intra-rater reliability (ICC = 0.79–0.84) [6], suggests that counting errors can happen and that these are likely to occur when administering the test consecutively for multiple participants (e.g., in a clinical setting). There is a need for objective approaches for determining the number of repeated rise-UTT movements that are completed and how many are completed over a 'full'/acceptable range, which could then be used in a range of testing environments. Inertial measurement units (IMU) may provide a pragmatic approach for such testing.

In a preliminary study we demonstrated how IMUs might be used to assess the UTT-30, by determining the number of UTT movements completed and by suggesting how movement consistency across the repeated UTT movements might be assessed [11]. The present study has two aims: (i) to determine whether IMUs provide a valid assessment of the UTT-30, by evaluating whether IMU-derived assessment metrics are comparable to those gained from a force-platform (FP); and (ii) to determine whether IMUs provide valid assessment of a single UTT movement when the participant is asked to hold the position of being UTT for 5 s (UTT-stand). For this (newly presented) UTT-stand test, we describe how IMUs could be used to assess the movement dynamics during the rise-UTT, as well as the postural control exhibited whilst holding the UTT position, and again evaluate whether IMU-derived assessment metrics are comparable to those gained from an FP. There are many daily living locomotor activities that involve being momentarily balanced on just the ball (toes region) of the foot e.g., during the single-limb support phase of stair descent when reaching downwards with the leading limb to contact the next (lower) step. Thus, use of the instrumented UTT-stand test to assess a person's ability to balance on their toes could provide relevant additional insights regarding ankle strength and function that are adjunct to those provided by the UTT-30.

2. Methods

2.1. Participant Recruitment

Twenty young, healthy adults (13 males, 7 females; 26.2 ± 7.7 years, 1.7 ± 0.1 m and 69.9 ± 16.1 kg) were recruited from the University of Bradford and its surrounding local area. Participants were excluded if they had any neurological, cardiovascular or musculoskeletal conditions that could affect their balance. All participants self-reported that they were lightly-to-vigorously physically active (as defined by the American College of Sports Medicine (ACSM) physical activity guidelines) [12]. The study received ethical approval, and all participants gave informed consent. Participants wore comfortable clothing and flat-soled shoes (trainers), and any corrective vision lens/spectacles that they habitually wore for walking.

2.2. Protocol

Participants stood on a floor-mounted FP (AMTI OR6-6, Watertown, MA, USA) in an upright position (arms by the sides, feet comfortably apart). They were asked to maintain their gaze at an eye-level visual target placed 5 m in front of them. The test proceeded as follows:

1. Twenty seconds of stationary standing.
2. a UTT-30. Here, participants were instructed to rise-UTT as high as comfortably possible and at a comfortable speed until indicated to stop (after 30 s), ensuring the heel contacted the ground between each movement. Note, all UTT movements were accepted for later analysis.
3. Twenty seconds of stationary standing.
4. a UTT-stand. Here, participants were instructed to rise-UTT as fast as comfortably possible and then hold this position until indicated to "return" to normal standing (after 5 s).
5. Twenty seconds of stationary standing.

A stopwatch (iPhone XS) was used to time each element. If a participant lost their balance during the UTT-stand, they were asked to repeat steps 4 and 5, above.

2.3. Data Collection

Ground reaction force (GRF) data (at 100 hertz, Hz) were collected for the entire UTT protocol i.e., steps 1–5 as described above. Data were low-pass filtered (Butterworth) with a cut-off of 20 Hz. The filtered GRF data in the vertical (Fz) direction along with the centre of pressure (CoP) coordinate data in the anteroposterior (CoPy) directions were exported in ASCII ('.csv') format for further analysis.

Two IMUs (AX6, Axivity Ltd., Newcastle, UK, https://axivity.com/ accessed on 3 May 2021: tri-axial accelerometer and tri-axial gyroscope) were attached to the upper surface of each shoe (i.e., on the shoelaces/tongue area). A third IMU (AX3, Axivity Ltd., tri-axial accelerometer) was attached over the lower back (fifth lumbar vertebrae, L5). The technologies have been previously evaluated and deemed suitable to capture data on human movement [13]. Using the proprietary software [14], IMUs (16-bit resolution, $\pm$8 g, $\pm$1000°/s) were set to capture data at 100 Hz. Prior to attachment, they were held together in the hands of the researcher (all oriented in the same direction) who then 'banged' their hands (with the IMUs)' on a desk. This introduced an acceleration spike in each IMU, which was used as a 'time stamp' to retrospectively synchronize/align data from each IMU in time. Once the UTT-tests were concluded, IMUs were removed, and data were downloaded for further analysis.

2.4. Data Treatment and Analysis

2.4.1. Forceplatform Data Analysis

Using CoPy (anteroposterior, AP) data, CoP AP velocity (CoPyVel) and CoP AP acceleration (CoPyAccel) were determined using a 4-point moving window differentiation approach. The following were then determined:

UTT-30 (Figure 1a): Num. UTT Movements Completed, and Mean Peak CoPyVel

The occurrence of a UTT movement was visually identified as a distinct peak (local maximum) in the CoPyVel. For each participant, the mean and standard deviation (SD) in peak CoPyVel across their repeated UTT movements were calculated, and any CoPyVel peaks with a magnitude less than 1 SD below the mean were highlighted. The total number of CoPyVel peaks identified indicated the number of UTT movements attempted, and the number of CoPyVel peaks within 1 SD of the mean, indicated the number of UTT movements that were 'fully' completed. Each participant's mean peak CoPyVel was then recalculated from the fully completed UTT movements.

In our previous (sister) study, we identified movements completed over a sufficiently 'full' range as those with a peak CoPyVel that were within 2 SD of the mean CoPyVel [11]. However, because identifying a peak in CoPyVel to identify each UTT movement attempted provides an indication of UTT speed but not of its ROM, we highlighted that instead of identifying (and discounting) UTT movements that are not completed over an acceptable/'full' ROM, an alternative approach would be to determine UTT consistency. Consistency could be evaluated simply by determining how many UTT movements, in percentage terms, had a peak CoPyVel that was within a certain threshold of the mean peak CoPyVel [11]. We also suggested consistency could be determined for a range of thresholds, e.g., within 1 SD, 1.5 SD and 2 SD of the mean. As the key aim in the present study was to determine the validity of the use of IMUs in comparison with the use of an FP, for pragmatic purposes we chose a 1 SD threshold to identify movements completed with an acceptable range. Such a threshold is the most widely used when assessing movement variability.

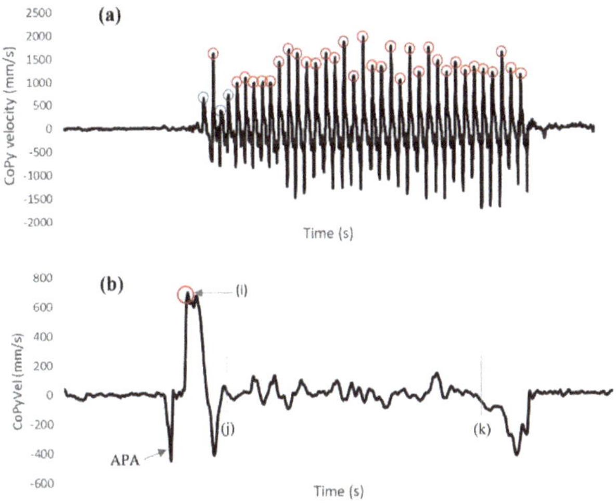

Figure 1. (**a**) Exemplar centre of pressure anteroposterior velocity (CoPyVel) trajectory showing identification of local maxima (circled in red) for each repeated up-on-the-toes (UTT) movement during the UTT-30 s test with peaks falling below 1 SD of the mean (circled in blue). (**b**) Exemplar CoPyVel trajectory from UTT-stand test, (i) indicates the local maximum in CoPyVel when rising UTT, (j) indicates the beginning of 'holding' the UTT position, and (k) indicates the end of 'holding' the UTT position. NB, the local minimum in CoPyVel prior to the local maximum indicates the anticipatory postural adjustment (APA) made when initiating the UTT movement.

UTT-Stand (Figure 1b): Peak CoPyVel, and Peak CoPy Accel (Figure 1b)

The UTT-stand was identified by a single distinct increase in the CoPyVel following an anticipatory postural adjustment (APA) that indicated the initiation of the UTT-stand. The APA was identified by a distinct negative peak in CoPyVel. The peak CoPyVel and peak CoPy Accel occurring during the rising UTT position were recorded.

When standing stationary, a force averaging one body weight (BW) is applied to the ground, and hence there will be a vertical (z) reaction force of 1 BW. Thus, having a vertical reaction force (Fz measured in Newtons) greater than BW applied over a time period, will create a vertical impulse that causes the centre of mass to rise upwards. The Fz impulse (above BW threshold) for rising UTT was calculated for the UTT-stand using the Simpson's Rule (Equation (1)).

$$\int_{Xo}^{Xn} f(x)dx = \frac{1}{2}h[\gamma_o + \gamma_n] + 2(\gamma_1+\gamma_2+\ldots+\gamma_{n-1}) \ (\mathrm{N_s}) \quad (1)$$

where h is the interval in time (0.01) and y is each Fz value from the instance Fz first rose above 1 BW for 5 samples/frames (o) to the instance the Fz fell back to 1 BW following the holding of the UTT position.

The instant that the UTT holding position was begun was determined as the instant that the CoPyVel reduced to less than 50 mm/s, following a brief increase in CoPyVel after it had become negative as participants moved to the UTT position. The instant that the holding of the UTT position was ended was determined as the instant CoPyVel first reduced below −50 mm/s prior to a negative peak in CoPyVel, as participants returned to normal standing. The postural stability when holding the UTT position was determined as the SD in the CoPyVel for the period that participants held the UTT position.

2.4.2. IMU Data Analysis

Custom programs in MATLAB® (2019, MathWorks, Inc., Natick, MA, USA: www.mathworks.com (accessed on 31 May 2023)) were used to automatically segment each UTT test from the continuous streams of IMU data and extract outcomes. Moving SD, mean and sum operations were initially used to shape the signal for use in segmentation of UTT tests. Figure 2a highlights the segmentation processes.

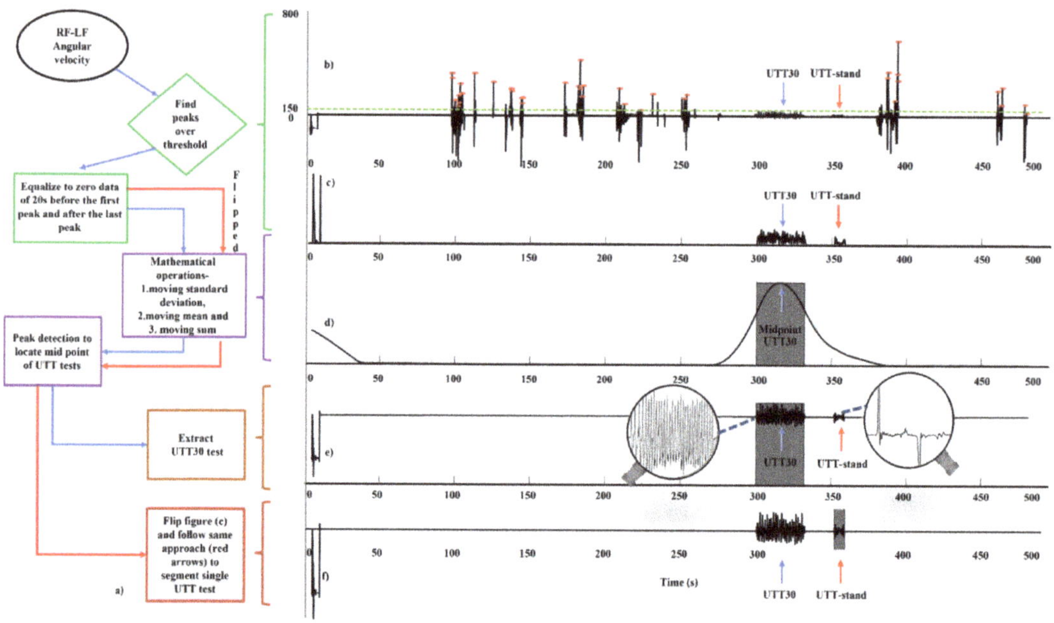

Figure 2. Automatic segmentation of up-on-the-toes 30 s test (UTT-30, blue arrows) and UTT-stand test (UTT-stand, red arrows) from right foot (RF) and left foot (LF) inertial measurement units (IMUs). (**a**) Flow chart of the segmentation algorithm; (**b**) raw angular velocity signal with specified peaks and tests; (**c**) processed angular velocity signal; (**d**) signal after mathematical operations; (**e**) extracted UTT-30; and (**f**) extracted UTT-stand.

Analysis was begun by identification/extraction of the UTT-30 data as it generated clear and repetitive angular velocity patterns for the right and left feet (RF and LF) in the sagittal plane during plantar flexion. From a visual examination of all participants' data, a threshold value of 150°/s was set to extract the UTT tests (Figure 2b, green dashed line). The automated segmentation algorithm is as follows:

1. A peak detection algorithm was used to detect the peaks (local maxima) above the 150°/s threshold.

2. Data for the 20 s before the first peak and after the last peak was equalized/set to zero, i.e., the algorithm requires and makes use of periods of 20 s stationary standing before and after the UTT-30 (Figure 2c).
3. Moving SD and moving mean operations with a 5 s sliding window were used to differentiate the UTT-30 from UTT-stand.
4. Moving summing operations along with peak detection was used to detect the mid-point of the repeated UTT-30 (Figure 2d). The sliding window size of the moving sum operation was set to 30 s (i.e., the length of UTT-30).
5. The UTT-30 was segmented considering 15 s before and after the calculated mid-point/peak location (Figure 2e).
6. Once the UTT-30 was segmented, the signal was flipped in the time domain (MATLAB® 'flipud' function), reversing the chronological order of the data. This allowed the UTT-stand to be segmented using the same approach as described in step 3 above (e.g., moving SD, Figure 2f), with the sliding window size of the moving sum operation set to 5 s (i.e., length of a UTT-stand).

Reversing the signal in time ensured the last (successful) UTT-stand was segmented if a participant had to repeat the test due to losing their balance in their first attempt. The automatic segmentation (Figure 2) was performed for LF and RF separately. Once the UTT tests were segmented, timestamp information of LF-RF IMUs was used to segment UTT tests from the AX3 sensors attached to L5.

Number of Movements, and Mean Peak Foot Angular Velocity

A UTT movement was identified as a distinct peak (local maximum) in RF and LF plantarflexion angular velocity (sagittal plane). The mean and SD peak angular velocity across the repeated UTT movements were calculated, and any peaks with a magnitude of less than 1 SD below the mean were highlighted. The total number of angular velocity peaks identified indicated the number of UTT movements attempted, and the number of peaks within 1 SD of the mean indicated how many UTT movements were completed 'fully'. (As highlighted above, a threshold of 1 SD below the mean was for pragmatic purposes.) Each participant's mean peak plantarflexion velocity was then recalculated from the fully completed UTT.

Peak Plantarflexion Angular Velocity

Each foot's maximum plantarflexion angular velocity (sagittal plane) during the rising-to-the-toes movement was extracted. From the values determined for both the RF and LF, the average value was recorded as peak plantarflexion angular velocity.

Peak CoM (L5) Upwards Acceleration (When Rising to Toes)

Data from the IMU attached to L5 were low-pass filtered (2 Hz, zero-lag 4th-order Butterworth) [13]. Then the peak acceleration in the vertical direction during the rise-to-toes movement was identified. This peak was recorded as peak body centre of mass (CoM) upwards acceleration.

Postural Stability Whilst Holding UTT Position

Postural stability was then assessed as the variability (SD) in the AP acceleration of the L5 IMU (CoM) whilst holding the UTT position. The holding position was deemed to start at 0.4 s after the peak in upward acceleration and to end 0.4 s before the peak in downward acceleration.

2.5. Analysis: Comparison between IMU and FP Outcomes

As parameters derived from the IMU and FP were determined from different (transducer) technologies, it would be inappropriate to determine 'agreement' between their outcome measures. However, it was expected that the measures derived would be related to some extent. For example, (i) to evaluate how quickly an individual rose to their toes during the UTT-stand, the peak CoPy velocity (FP) and the peak foot angular velocity (IMU) were determined and thus one would expect the two different measures would be correlated and; (ii) the peak foot angular velocity during rising to the toes might also be associated with how much 'push' is developed in raising the whole-body CoM. Accordingly, one might expect that this measure would also be closely related to the Fz-impulse generating when rising up to the toes. Therefore, to highlight the relationship between IMU and FP measures, we created a series of scatter (x, y) graphs (using data for all participants), plotting each outcome measure determined from the IMU and corresponding FP measure. Each plot contains a linear regression line where the r-value indicates the strength of the relationship (correlation) between IMU and FP measures. R-values <0.35 were considered to indicate a low-to-weak relationship between IMU and FP measures; 0.36–0.67 moderate; and ≥0.68 as strong [15].

For the UTT-30 test, the following graphs were plotted:

- number of fully completed UTT movements determined by IMU against number determined by FP (Figure 4a).
- the mean peak plantarflexion angular velocity against the mean peak CoPy velocity (Figure 4b).

For the UTT-stand, the following graphs were plotted:

- for the period when moving UTT: peak plantarflexion angular velocity against the maximum CoPy velocity (Figure 5a); the peak L5 upward acceleration against the maximum CoPy acceleration (Figure 5b); and the peak plantarflexion angular velocity against the vertical (Fz) impulse (Figure 5c).
- for the period when holding the UTT position: the SD in L5 horizontal acceleration against the SD in CoPy velocity (Figure 5d).

3. Results

All 20 participants completed the UTT tests and showed no signs of excessive discomfort, fatigue, or pain during the assessments. Figure 3 shows an example of the raw (sample level) IMU and FP data.

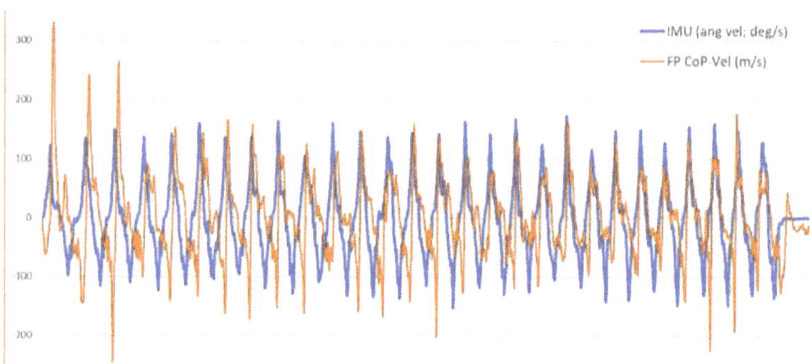

Figure 3. Example of raw (example level) inertial measurement unit (IMU, angular velocity, blue line) and force-platform (FP) centre of pressure (CoP, orange line) velocity (m/s) signals from an up-on-the-toes 30 s test (UTT-30).

3.1. UTT-30

3.1.1. Number of Movements

The group average number of movements (full or partial) detected by IMU (23.5) and FP (23.6), matched the number counted by an observer (23.6) (Table 1). After elimination of any detected movements not completed within 1 SD of each participant's mean peak angular velocity (IMU) or mean peak CoPyVel (FP), the number of movements determined decreased to 19.7 (IMU) and 19.9 (FP), respectively (Table 1), indicating a strong correlation between assessment approaches (r = 0.99, Figure 4a). The mean difference between the IMU approach and the FP approach, in determining the number of UTT movements fully completed, was 0.25, with a 95% agreements limit of −3.7–4.2.

Table 1. The number of up-on-the-toes (UTT) movements completed by each participant as determined by an observer, and by analysis of the force-platform (FP) or inertial measurement units (IMUs) data and each participant's mean peak centre of pressure anteroposterior (CoPy) velocity and mean plantarflexion (p-f) angular velocity (for fully completed UTT movements).

Participant	No. UTT Counted by Observer	No of UTT Movements Detected	No of UTT Movements Detected	No. UTT Fully Completed	No. UTT Fully Completed	Mean Peak CoPyVel (mm/s)	Mean Peak p-f Angular Velocity (°/sec)
		FP	IMU	FP	IMU	FP	IMU
1	29	29	29	22	24	613.0	169.1
2	36	36	36	33	30	957.1	138.4
3	12	12	12	11	9	932.0	169.7
4	24	24	24	22	22	563.4	159.9
5	36	36	36	27	32	980.6	272.7
6	31	31	31	27	27	456.0	193.4
7	19	19	19	16	17	663.2	189.7
8	28	28	28	26	22	559.7	186.8
9	21	21	21	17	16	741.6	163.1
10	23	23	23	21	20	388.2	140.5
11	26	26	26	21	23	1116.6	183.9
12	18	18	18	15	15	581.7	172.4
13	17	17	17	14	14	647.3	105.5
14	12	12	12	11	9	358.5	83.1
15	18	18	18	15	15	485.6	131.5
16	15	15	15	12	13	503.0	106.9
17	18	18	18	15	16	724.0	157.1
18	17	17	16	15	13	555.4	74.3
19	19	19	19	15	14	908.5	149.9
20	52	52	52	43	42	621.9	174.7
Mean ± SD	23.6 ± 9.7	23.6 ± 9.7	23.5 ± 9.8	19.9 ± 8.1	19.7 ± 8.3		

3.1.2. CoPy Velocity and Plantarflexion Angular Velocity

The strength of the relationship between participants' mean peak CoPy velocity (FP) and their mean peak plantarflexion angular velocity (IMU), was moderate (r = 0.460, Figure 4b).

3.2. *UTT-Stand*

3.2.1. Period When Rising to the Toes

The strength of the relationship between participants' plantarflexion angular velocity (IMU) and the maximum CoPy velocity (FP) or the Fz impulse (FP) was strong (r = 0.745 and r = 0.805 respectively, Figure 5a,c), whereas the strength of the relationship between peak L5 upward acceleration (IMU) and the maximum CoPy acceleration (FP) was moderate (r = 0.596, Figure 5b).

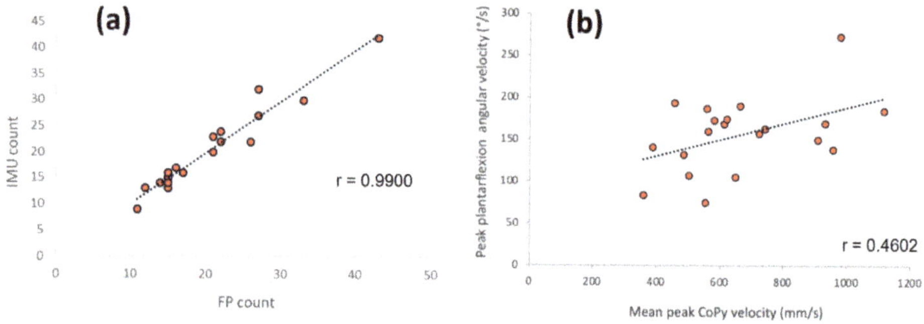

Figure 4. (**a**) The number of up-on-the-toes (UTT) movements each participant completed fully during the UTT-30 test determined using inertial measurement units (IMUs) plotted against a number determined using forceplatform (FP). (**b**) Each participant's mean peak plantarflexion angular velocity (from IMU) plotted against their mean peak centre of pressure anteroposterior (CoPy) velocity values (from FP) for the UTT-30 test. 'r' is Pearson correlation coefficient. Dotted line in each plot indicates the linear regression line.

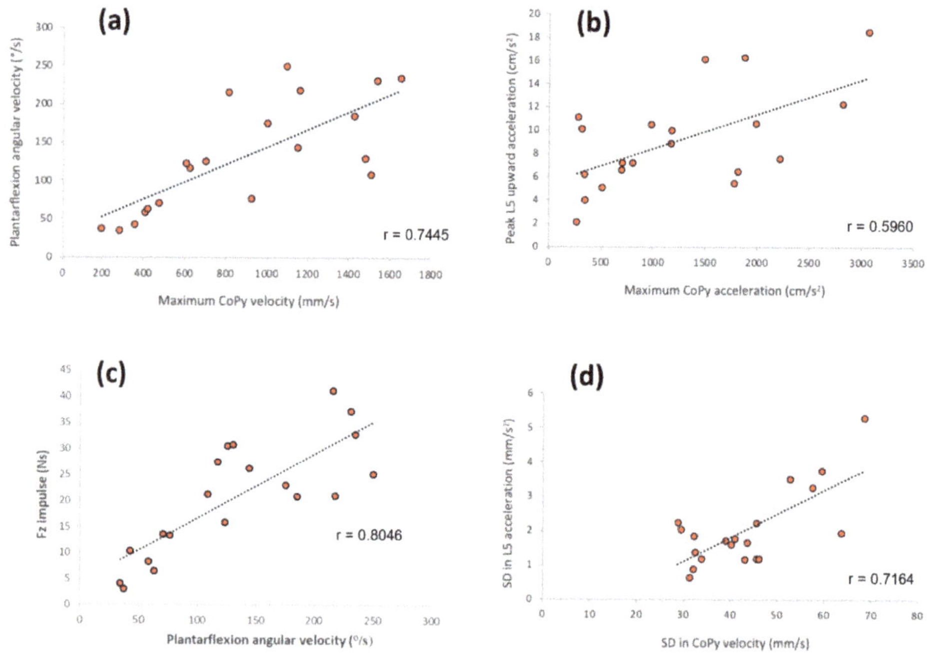

Figure 5. Outcomes for the up-on-the-toes (UTT) stand test: inertial measurement unit (IMU)-derived measures plotted against forceplatform (FP)-derived measures for the period when moving onto the toes. (**a**) Peak plantarflexion angular velocity (from IMU) plotted against peak centre of pressure anteroposterior (CoPy) velocity (from FP); (**b**) peak 5th lumbar vertebrae (L5) upward acceleration (from IMU) plotted against peak CoPy acceleration (from FP); (**c**) peak plantarflexion angular velocity (from IMU) plotted against the vertical ground reaction force (Fz) impulse (from FP) and the period when holding the UTT position; and (**d**) standard deviation (SD) in L5 acceleration (IMU) plotted against SD in CoPy velocity (FP). 'r' is Pearson correlation coefficient. Dotted line in each plot indicates the linear regression line.

3.2.2. Period of Holding the UTT Position

The strength of the relationship between participants' SD in L5 acceleration (IMU) and the SD in CoPy velocity (FP) was strong (r = 0.716, Figure 5d).

4. Discussion

The aims of the present study were to (i) determine if IMUs can provide valid assessment of the UTT-30 test, by evaluating whether IMU-derived assessment metrics are comparable to those gained from a force-platform, and (ii) to determine whether IMUs provide valid assessment of a single UTT movement when the participant is asked to hold the UTT position for 5 s (UTT-stand). For this (newly presented) UTT-stand test, we describe how IMUs could be used to assess the movement dynamics during the rise-UTT, as well as the postural control exhibited whilst holding the UTT position, and again evaluate whether IMU-derived assessment metrics are comparable to those gained from a FP. Our findings highlight that the IMU automated approach determined the same number of repeated UTT movements completed as that determined by an FP (UTT-30) and yielded movement and dynamic postural control measures that were comparable (i.e., had moderate-to-strong correlation) to those determined from using an FP (UTT-stand).

There is considerable recent work that has reported ways in which IMUs might be used for the assessment of standing postural stability and in the assessment of gait outcomes across a wide range of research disciplines [16–23]. In our sister study we demonstrated how IMUs might be used to assess the UTT-30 [11]. The current study extends upon this recent work by showing that IMU-derived assessment metrics are comparable to those gained from a force-platform, and thus highlights how IMUs might provide valid assessment of a UTT-30 test. It also extends this previous work by highlighting how IMUs can be used to assess a UTT-stand test. This newly presented test provides relevant additional insights of ankle strength and function that are adjunct to those provided by the UTT-30 test.

4.1. UTT-30

Results show that the IMUs detected the same number of UTT attempted movements as that determined by the FP (and counted by an observer). Furthermore, the IMUs and FP both provided an objective means to discount any UTT movements that were deemed to have not been completed 'fully'. UTT movements that were not completed 'fully' were those that yielded a peak velocity value during the period of rising to up-on-the-toes that was less than 1 SD below each participant's mean peak velocity (either plantarflexion velocity, IMU, or CoP AP-velocity, FP) of rising to up-on-the-toes for all their attempted repeated UTT movements. Identifying UTT movements that might not be completed properly ('fully') would be difficult to do by 'the naked eye', because an observer would only be able to count how many UTT movements were attempted. Both the IMUs and FP assessment approaches indicated that the average number of UTT movements completed 'fully' was 19.7–19.9 (±8; indicating a strong correlation between assessment approaches, r = 0.99), and this was, on average, 3.8 fewer than the number of UTT attempted movements.

As identifying a peak in CoPyVel to identify each UTT movement attempted provides indication of UTT speed but not its ROM, we highlighted in our previous study [11] that instead of trying to identify (and discount) UTT movements that are not completed over a sufficiently 'full' range, perhaps a better approach would be to determine UTT consistency. Consistency could be evaluated simply by determining how many UTT movements, in percentage terms, had a peak CoPyVel that was within a certain threshold of the mean peak CoPyVel [11]. We suggested that the threshold could be within 1 SD, 1.5 SD or 2 SD of the mean. UTT consistency may have more clinical relevance than determining the number of UTT movements 'fully' completed, though future work is needed to confirm this.

Each participant's mean peak plantarflexion velocity, derived from the IMUs, for the 'fully' completed repeated UTT movements reflect how quickly, on average, an individual was rising up to their toes during the 30 s test period. The association between mean peak plantarflexion velocity and the mean peak CoPyVel was determined via regression analysis, which showed the strength of the relationship between the two outcome measures (r = 0.46). Having only a moderate strength relationship is likely due to the IMU signal representing local changes at the ankle as opposed to the FP signal representing the speed of movement of the whole-body CoM.

4.2. UTT-Stand

Our assessment focused on determining the velocity an individual rises up to their toes, and the postural stability of holding the UTT position. The results show the strength of the relationships between IMU-derived and FP-derived measures were moderate-to-strong for speed of rising onto the toes, and for the postural stability exhibited whilst holding the UTT position (Figure 5d). The UTT-stand test assesses a person's ability to rise onto the toes and then to maintain postural stability when holding the UTT position. As such, it provides a complementary assessment of ankle function to the UTT-30 test. As this is a newly presented test, future work is needed to determine if and how outcomes from the UTT-stand test are related to traditional measures of ankle strength and function.

4.3. Limitations and Recommendations for Future Work

When analysing the L5 IMU UTT-stand data of two participants, our algorithm performed poorly in detecting any prominent acceleration peaks depicting when the participants started rising up on to their toes. Although there were acceleration peaks present, the amplitudes were much lower than expected (by the algorithm), which may have been due to these two participants performing the UTT movement much slower than the other participants. As the participants in the current study were all healthy, the two participants that performed the UTT movement much slower may have misunderstood instructions. This highlights further that work is needed to determine how the segmentation/analysis algorithm (used to analysis IMU data) could be automatically tailored when used, for example, in elderly, frail individuals who will likely perform UTT movements at a slower speed and with more variability.

Previous work has shown that ankle strength and function is related to falls risk [24–26]. Thus, further work should also determine if the measures from the UTT-30 and the UTT-stand are related to a person's physical functioning (e.g., strength/endurance of ankle musculature) and/or a person's falls history/ fall risk.

Although use of FP might be considered a gold standard method for measuring postural control during standing, reaching and rising to the toes, IMUs are versatile in assessing aspects of postural control due to their ability to be placed almost anywhere on the body. For instance, to investigate stability during balance tasks, an IMU can be placed on the trunk or hip to capture acceleration data. However, as the AX6s used in the current study contain integrated gyroscopes for angular velocity measurement, further measures such as angular displacement of the IMUs, i.e., segment-tilt angle, can also be derived. Hence, using these devices can provide complementary data/insights to those gained from the FP, in so far as acceleration of specific reference points and/or angular velocity of specific segments can be measured directly. Indeed, such direct measures can be undertaken beyond the lab in measuring physical functioning/activity in clinical tests or in free-living situations.

5. Conclusions

For both the UTT-30 and UTT-stand tests, IMU derived outcome metrics were comparable to FP derived metrics. This suggests that IMUs can provide a valid means to assess the UTT-30 and UTT-stand tests. Future work is needed to determine the most appropriate approach to assess UTT-30 performance using IMU-derived outcomes, and whether outcomes of the newly presented UTT-stand are associated with clinical outcomes.

Author Contributions: Conceptualization, S.A.Z., Y.C., A.G. and J.G.B.; methodology, Y.C.; software, Y.C.; validation, S.A.Z., Y.C., and J.G.B.; formal analysis, S.A.Z., Y.C., A.G. and J.G.B.; investigation, S.A.Z., Y.C., A.G. and J.G.B.; resources, S.A.Z., Y.C. and J.G.B.; data curation, S.A.Z.; writing—original draft preparation, S.A.Z.; writing—review and editing, S.A.Z., Y.C., A.G. and J.G.B.; visualization, Y.C. and J.G.B.; supervision, A.G. and J.G.B.; project administration, J.G.B. All authors have read and agreed to the published version of the manuscript.

Funding: This research received no external funding.

Institutional Review Board Statement: The study was conducted in accordance with the Declaration of Helsinki and Ethical approval was granted from University of Bradford's Committee for Ethics in Research: approval reference E.119.

Informed Consent Statement: Informed consent was obtained from all subjects involved in the study.

Data Availability Statement: Data will be made available upon reasonable request to the corresponding author.

Acknowledgments: The authors would like to thank all those who volunteered for this study. Yunus Celik receives PhD studentship support from the Turkish Ministry of National Education.

Conflicts of Interest: The authors declare no conflict of interest.

References

1. Pol, F.; Baharlouei, H.; Taheri, A.; Menz, H.B.; Forghany, S. Foot and ankle biomechanics during walking in older adults: A systematic review and meta-analysis of observational studies. *Gait Posture* **2021**, *1*, 14–24. [CrossRef] [PubMed]
2. Spink, M.J.; Fotoohabadi, M.R.; Wee, E.; Hill, K.D.; Lord, S.R.; Menz, H.B. Foot and ankle strength, range of motion, posture, and deformity are associated with balance and functional ability in older adults. *Arch. Phys. Med. Rehabil.* **2011**, *1*, 68–75. [CrossRef]
3. Menz, H.B.; Morris, M.E.; Lord, S.R. Foot and Ankle Characteristics Associated with Impaired Balance and Functional Ability in Older People. *J. Gerontol. Ser. A* **2005**, *60*, 1546–1552. [CrossRef]
4. Buckley, J.G.; Cooper, G.; Maganaris, C.N.; Reeves, N.D. Is stair descent in the elderly associated with periods of high centre of mass downward accelerations? *Exp. Gerontol.* **2013**, *48*, 283–289. [CrossRef] [PubMed]
5. Riener, R.; Rabuffetti, M.; Frigo, C. Stair ascent and descent at different inclinations. *Gait Posture* **2002**, *15*, 32–44. [CrossRef]
6. André, H.I.; Carnide, F.; Borja, E.; Ramalho, F.; Santos-Rocha, R.; Veloso, A.P. Calf-raise senior: A new test for assessment of plantar flexor muscle strength in older adults: Protocol, validity, and reliability. *Clin. Interv. Aging* **2016**, *11*, 1661–1674. [CrossRef]
7. Hébert-Losier, K.; Newsham-West, R.J.; Schneiders, A.G.; Sullivan, S.J. Raising the standards of the calf-raise test: A systematic review. *J. Sci. Med. Sport* **2009**, *12*, 594–602. [CrossRef]
8. Jan, M.-H.; Chai, H.M.; Lin, Y.F.; Lin, J.C.; Tsai, L.Y.; Ou, Y.C.; Lin, D.H. Effects of age and sex on the results of an ankle plantar-flexor manual muscle test. *Phys. Ther.* **2005**, *85*, 1078–1084. [CrossRef]
9. Lunsford, B.R.; Perry, J. The standing heel-rise test for ankle plantar flexion: Criterion for normal. *Phys. Ther.* **1995**, *75*, 694–698. [CrossRef] [PubMed]
10. Sman, A.D.; Hiller, C.E.; Imer, A.; Ocsing, A.; Burns, J.; Refshauge, K.M. Design and reliability of a novel heel rise test measuring device for plantarflexion endurance. *BioMed Res. Int.* **2014**, *2014*, 391646. [CrossRef]
11. Zahid, S.A.; Celik, Y.; Godfrey, A.; Buckley, J.G. Use of 'wearables' to assess the Up-on-the-toes test. *J. Biomech.* **2022**, *143*, 111272. [CrossRef]
12. American College of Sports Medicine. *ACSM's Guidelines for Exercise Testing and Prescription*; Lippincott Williams & Wilkins: Philadelphia, PA, USA, 2013.
13. Ladha, C.; Jackson, D.; Ladha, K.; Nappey, T.; Olivier, P. Shaker table validation of OpenMovement AX3 accelerometer. In Proceedings of the Ahmerst (ICAMPAM 2013 AMHERST): 3rd International Conference on Ambulatory Monitoring of Physical Activity and Movement, Amherst, MA, USA, 17–19 June 2013.
14. Github DigitalInteraction. *AX3 GUI*. Available online: https://github.com/digitalinteraction/openmovement/wiki/AX3-GUI (accessed on 2 December 2019).
15. Taylor, R. Interpretation of the Correlation Coefficient: A Basic Review. *J. Diagn. Med. Sonogr.* **1990**, *6*, 35–39. [CrossRef]
16. O'Brien, M.K.; Hidalgo-Araya, M.D.; Mummidisetty, C.K.; Vallery, H.; Ghaffari, R.; Rogers, J.A.; Lieber, R.; Jayaraman, A. Augmenting clinical outcome measures of gait and balance with a single inertial sensor in age-ranged healthy adults. *Sensors* **2019**, *19*, 4537. [CrossRef] [PubMed]
17. Hasegawa, N.; Maas, K.C.; Shah, V.V.; Carlson-Kuhta, P.; Nutt, J.G.; Horak, F.B.; Asaka, T.; Mancini, M. Functional limits of stability and standing balance in people with Parkinson's disease with and without freezing of gait using wearable sensors. *Gait Posture* **2021**, *87*, 123–129. [CrossRef] [PubMed]
18. Kang, G.E.; Stout, A.; Waldon, K.; Kang, S.; Killeen, A.L.; Crisologo, P.A.; Siah, M.; Jupiter, D.; Najafi, B.; Vaziri, A.; et al. Digital Biomarkers of Gait and Balance in Diabetic Foot, Measurable by Wearable Inertial Measurement Units: A Mini Review. *Sensors* **2022**, *22*, 9278. [CrossRef] [PubMed]

19. Noamani, A.; Nazarahari, M.; Lewicke, J.; Vette, A.H.; Rouhani, H. Validity of using wearable inertial sensors for assessing the dynamics of standing balance. *Med. Eng. Phys.* **2020**, *77*, 53–59. [CrossRef]
20. Song, S.; Nordin, A.D. Balance Perturbations in Simulated Low-Gravity Modulate Human Premotor and Frontoparietal Electrocortical Theta, Alpha, and Beta Band Spectral Power. *IEEE Open J. Eng. Med. Biol.* **2023**, 1–9. [CrossRef]
21. Young, F.; Coulby, G.; Watson, I.; Downs, C.; Stuart, S.; Godfrey, A. Just Find It: The Mymo Approach to Recommend Running Shoes. *IEEE Access* **2020**, *8*, 109791–109800. [CrossRef]
22. Celik, Y.; Powell, D.; Woo, W.L.; Stuart, S.; Godfrey, A. A feasibility study towards instrumentation of the Sport Concussion Assessment Tool (iSCAT). In Proceedings of the 2020 42nd Annual International Conference of the IEEE Engineering in Medicine & Biology Society (EMBC), Montreal, QC, Canada, 20–24 July 2020.
23. Furtado, S.; Godfrey, A.; Del Din, S.; Rochester, L.; Gerrand, C. Are Accelerometer-based Functional Outcome Assessments Feasible and Valid After Treatment for Lower Extremity Sarcomas? *Clin. Orthop. Relat. Res.* **2020**, *478*, 482–503. [CrossRef]
24. Cattagni, T.; Scaglioni, G.; Laroche, D.; Grémeaux, V.; Martin, A. The involvement of ankle muscles in maintaining balance in the upright posture is higher in elderly fallers. *Exp. Gerontol.* **2016**, *1*, 38–45. [CrossRef]
25. Menz, H.B.; Morris, M.E.; Lord, S.R. Foot and Ankle Risk Factors for Falls in Older People: A Prospective Study. *J. Gerontol. Ser. A* **2006**, *61*, 866–870. [CrossRef] [PubMed]
26. Cattagni, T.; Scaglioni, G.; Laroche, D.; Van Hoecke, J.; Gremeaux, V.; Martin, A. Ankle muscle strength discriminates fallers from non-fallers. *Front. Aging Neurosci.* **2014**, *6*, 336. [CrossRef] [PubMed]

biomechanics

MDPI

Article

Mimicking an Asymmetrically Walking Visual Cue Alters Gait Symmetry in Healthy Adults

Krista G. Clark [1], Louisa D. Raisbeck [1], Scott E. Ross [1] and Christopher K. Rhea [2,*]

1 Department of Kinesiology, University of North Carolina Greensboro, Greensboro, NC 27412, USA; kgmeder@uncg.edu (K.G.C.); ldraisbe@uncg.edu (L.D.R.); seross@uncg.edu (S.E.R.)
2 College of Health Sciences, Old Dominion University, Norfolk, VA 23529, USA
* Correspondence: crhea@odu.edu

Abstract: Gait asymmetries are a common problem in clinical populations, such as those with a history of stroke or Parkinson's disease. The use of a split-belt treadmill is one way to enhance gait symmetry but relies on specialty (and typically expensive) equipment. Alternatively, visual cues have been shown as a method to alter gait mechanics, but their utility in altering gait symmetry has been relatively understudied. Before deploying this method to clinical populations, a proof-of-concept study is needed to explore using visual cues to alter gait symmetry in healthy adults. Therefore, the purpose of this study was to examine the extent to which healthy adults could synchronize to an asymmetric visual cue with a small or large gait asymmetry using wearable sensors to measure gait asymmetries. Seventy-two healthy adults (ages: 23.89 ± 6.08 years) walked on the treadmill for two conditions: with and without the visual cue. Each walking condition lasted 10 min at the participant's preferred walking speed. Inertial sensors were used to measure gait asymmetries. Some participants did not respond to the visual cue, and groups were separated into responders and non-responders. Participants in the small and large asymmetry-responder groups exhibited statistically significant increased asymmetries in single limb support % ($p < 0.01$) and step duration (s) ($p < 0.05$, $p < 0.01$, respectively). Only the large asymmetry-responder group showed statistically significant ($p < 0.01$) increased asymmetries in stride length. Overall, asymmetrical walking visual cues can alter gait asymmetries, and inertial sensors were sensitive enough to detect small changes in gait asymmetries.

Keywords: gait; asymmetry; symmetry; inertial sensors; wearables; visual cue

1. Introduction

Neurological populations—such as people with a history of stroke or Parkinson's disease—commonly exhibit mobility impairments inclusive of gait asymmetries. People with gait asymmetries reduce the amount of weight bearing on the paretic or injured limb, which can lead to further muscular imbalances and a loss of bone mineral density [1]. Therefore, a primary goal of gait rehabilitation is to improve weight bearing on the paretic or affected limb [2]. In clinical and elderly populations, cognitive demand can impact the magnitude of gait asymmetries, as exhibited by dual tasks leading to increased gait asymmetries [3]. Gait asymmetries are also associated with a decreased walking speed and balance control and an increased metabolic cost and fall risk [2,4–6]. Falls can lead to a reduced quality of life, critical injuries, and loss of independence, all of which are of great concern for clinical and aging populations who exhibit gait asymmetries [7–13]. Thus, gait asymmetries are important to assess in clinical and aging populations and are a better indicator of functional recovery than gait velocity due to the weight-bearing aspect [2,14].

Methods to reduce gait asymmetries and fall risk have heavily relied on split-belt treadmill training in clinical populations [15–19]. The split-belt treadmill training paradigm requires the participant to move each limb at a different speed due to the two treadmill belts (one under each limb) moving at different rates. This method can be used as an error-based motor learning strategy during an adaptation (i.e., training) phase to reduce gait asymmetries when the belts are once again moving at the same speed. Split-belt treadmill training has shown success in altering gait asymmetries (e.g., step length, stride length, and double support time asymmetries) during adaptation conditions in which the belts are split in young healthy adults and in clinical populations [16,18,20]. Aftereffects (i.e., changes to the asymmetry remained after the belts went from split to tied) have been observed in step length in young healthy adults and clinical populations, as well as for double support time [16,18]. However, some gait metrics like stance time have been shown to be more difficult to alter with split-belt treadmill training [5,16,17]. Multiple sessions of split-belt treadmill training can improve step length asymmetries, and the improvements last from one to several months post training [17,21]. Thus, the data show that split-belt treadmill training can reduce certain gait asymmetries in clinical populations. However, split-belt treadmills are expensive and mainly conducive to lab-based training only. Therefore, it is important to not only explore alternative cost-effective options to alter gait but also ways to obtain greater clinical applicability. One way to improve clinical applicability is through the use of wearable technology such as inertial sensors. Inertial sensors are a simpler and a more affordable option than split-belt treadmill training, which typically uses embedded force plates or motion capture technology to quantify gait asymmetries. Also, a major advantage is that inertial sensors can be used anywhere, and they produce reliable and sensitive data in healthy and clinical populations [22–24].

Besides split-belt treadmill training, gait can be altered through a variety of ways such as using external cueing through visual cues. Visual cueing (i.e., prescribing the desired gait mechanics with a visual) can be performed via projections on a screen or in immersive virtual reality, which have produced greater improvements in gait and gait asymmetries than traditional treadmill training alone [25–27]. Synchronization of movement can also occur with external cues such as another human, avatars, and auditory metronomes [28–31]. Synchronization between two individuals can be intentional (i.e., cued) or spontaneous [32], and intentional synchronization can produce greater synchrony than spontaneous synchronization. Intentionally synchronizing to a human and an avatar in an immersive environment produces similar results, suggesting both methods are effective [31]. Thus, visual cueing combined with an additional external cue—such as intentional instructions to synchronize to a human or avatar—may be a way to alter gait asymmetry. However, the ability to synchronize with an asymmetric walking pattern that simulates training with a split-belt treadmill to alter gait symmetry is unknown. Before deploying this method to clinical populations, it is first important to show proof-of-concept that visual cueing can be used to alter gait symmetry in healthy adults. Given that healthy adults have relatively symmetrical gait patterns, they are an ideal initial population to test the feasibility and efficacy of interventions designed to induce gait asymmetry. In this preclinical work, we elected to first see if the gait patterns of healthy adults could be shifted toward asymmetry using an asymmetrically walking visual cue. If so, that would lay the foundation to test the reverse directionality in future research (i.e., can a symmetrically walking visual cue be used in asymmetrically walking clinical populations to develop more symmetrical gait patterns).

Therefore, the purpose of this study was to examine the extent to which healthy adults could synchronize to an asymmetric visual cue during a 10-min treadmill walking session with a small or large gait asymmetry using wearable sensors to measure gait asymmetries. This was part of a larger study that also examined the retention and transfer of gait asymmetries after the visual cue was removed. However, this manuscript specifically focuses on adaptation due to the novel methodological approach. We hypothesized that gait asymmetries would increase from the baseline to the adaptation phase, and the large asymmetry cue would induce greater gait asymmetries than the small asymmetry cue.

2. Materials and Methods

2.1. Participants

A total of 72 participants (49 females and 23 males) were recruited for this study and were quasi randomized into three groups: Small Gait Asymmetry (n = 32), who were given the visual cue with a 1.4–1 gait asymmetry ratio; Large Gait Asymmetry (n = 32), who were given the visual cue with a 1.9–1 gait asymmetry ratio; and Control (n = 8), who were given a non-stimulating nature documentary. The Small and Large Gait Asymmetry groups were later separated into responder and non-responder groups (Table 1). Our study design was primarily focused on the asymmetrically walking visual cue; however, the small control group served to ensure that our specific visual cue was what influenced changes in gait symmetry. Participants were between the ages of 18–50 years old, had a of BMI < 30, had normal or corrected-to-normal vision, and had the ability to walk continuously for at least 10 min. All participants reported no lower extremity injuries in the last 12 months and no musculoskeletal, cognitive, or neurological conditions.

Table 1. Participant demographics—means (SE).

Variables	Small Res (n = 20)	Small Non-R (n = 12)	Large Res (n = 22)	Large Non-R (n = 10)	Control (n = 8)	p
Age (years)	24.09 (6.29)	22.62 (7.30)	23.57 (4.33)	23.93 (6.98)	26.16 (2.37)	0.520
Height (cm)	168.29 (10.50)	166.32 (8.93)	167.97 (8.90)	169.20 (5.79)	168.20 (3.77)	0.822
Weight (kg)	70.22 (15.45)	68.00 (17.40)	71.07 (16.68)	75.89 (9.98)	69.97 (2.79)	0.564
Preferred walking speed (m/s)	0.88 (0.16)	0.93 (0.18)	0.84 (0.16)	0.96 (0.15)	1.14 (0.09)	0.011 *
Left leg length (cm)	88.40 (6.67)	88.31 (5.22)	87.69 (5.96)	89.32 (3.35)	88.60 (2.10)	0.860
Right leg length (cm)	88.37 (6.91)	88.75 (5.38)	87.72 (6.26)	89.12 (3.31)	88.68 (2.13)	0.906
Leg length difference (cm)	0.51 (0.30)	0.56 (0.44)	0.39 (0.30)	0.50 (0.28)	0.30 (0.07)	0.128

Note: Small Res = Small Asymmetry-Group Responders, Small Non-R = Small Asymmetry-Group Non-Responders, Large Res = Large Asymmetry-Group Responders, Large Non-R = Large Asymmetry-Group Non-Responders; * indicates $p < 0.05$.

2.2. Procedures

All data collection occurred in the Virtual Environment for Assessment and Rehabilitation (VEAR) laboratory at the University of North Carolina Greensboro. Prior to data collection, this study's procedures were approved by the institutional review board of the University of North Carolina Greensboro (IRB#: IRB-FY23-72). Participants completed the informed consent on Qualtrics and then were screened for limb length discrepancies via the direct method (i.e., measurements between the bony landmarks of the anterior superior iliac spine and the medial malleolus with a tape measure while laying supine) to ensure any discrepancies were less than 2 cm, as greater than 2 cm is related to gait asymmetries [33,34]. Participants completed a demographic/health history questionnaire, Footedness questionnaire, and the International Physical Activity Questionnaire short form (IPAQ-SF) via Qualtrics. Next, participants stepped onto the treadmill (Simbex Active Step, Lebanon, NH, USA) to determine their preferred walking speed [35]. Then, a total of seven Opal inertial sensors (APDM Inc., Portland, OR, USA) were placed on the following locations: the top and center of their shoes; the shins, with straps above the widest part of the gastrocnemius muscles; the lower lateral sides of the thighs at the midline; and the low back at the base of the spine. The sensors were calibrated prior to data collection, and the sensor sampling frequency was 128 Hz. Inertial sensors are reliable and are sensitive to measures in healthy and clinical populations [22–24]. Also, inertial sensors were used in this study to improve translatability to field-based studies.

Next, participants walked for two conditions at their preferred walking speed. Condition (1) walking for 10 min (baseline) with no visual cue, and condition (2) participants walked with the visual cue (adaptation) (Figure 1). The asymmetrically walking visual cue was projected on a screen in front of the treadmill. During adaptation, participants in the asymmetry groups were instructed to synchronize their gait to the visual cue via the following instructions "mimic the walking cue as closely as you can to the best of your ability." The control group were given the non-stimulating documentary video to watch during the adaptation condition and were instructed to walk normally. A five-minute seated rest period was provided between walking conditions to reduce the possibility of fatigue.

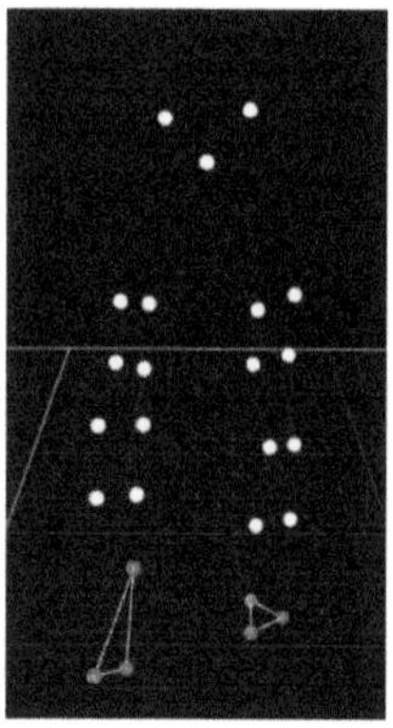

Figure 1. Freeze frame of the asymmetrically walking visual cue that shows the pelvis, legs, and feet that presented the motion of asymmetrical walking on a screen in front of the treadmill. Please see the Supplmentary Materials section for videos of the cues in motion.

The asymmetrical visual cue was created using data of healthy adults walking on a split-belt treadmill with varying belt speeds to create an asymmetrical walking pattern. To incorporate simulating a clinically relevant gait asymmetry, we used two different ratios of gait asymmetry for the asymmetrically walking visual cue based on the velocity of the treadmill belts. One visual cue has a small gait asymmetry with a ratio of 1.4–1 (i.e., the left side is moving 40% faster than the right). For example, the left belt has a velocity of 1.4 m/s and right belt has a velocity of 1.0 m/s). The other visual cue has a large gait asymmetry ratio of 1.9–1 (i.e., the left side is moving 90% faster than the right side). The visual cues have an average treadmill velocity of 1.24 m/s. The small asymmetrical ratio is slightly under the threshold of what is clinically detected based on observation alone [2], and the large asymmetrical ratio is close to a 2–1 ratio, which is commonly used in split-belt treadmill training to reduce gait asymmetries [16,17,36]. The dependent variables are single limb support (SLS) %, step duration (s), and stride length (m), given many gait asymmetry interventions focus on spatiotemporal measures [19]. The variables were calculated using the APDM provided software [37].

2.3. Statistical Analysis

Power calculations were performed a priori using G*Power 3.1 [38]. The parameter settings, using a *t*-test, were $\alpha = 0.05$, power = 0.8, and an effect size of 0.3. All data analyses were performed using R software (version 4.2.3) [39]. One-way ANOVAs were used to compare group demographics (Table 1). The Symmetry Index (SI) equation—a common method to quantify gait asymmetry [19,40]—was calculated for all dependent variables in R (Equation (1)). Gait asymmetries were considered present if the value was not zero [40].

$$\text{Symmetry Index (SI)} = \frac{x_L - x_R}{0.5 * (x_L + x_R)} * 100 \quad (1)$$

After visually inspecting the data, some participants exhibited minimal changes, while others exhibited more obvious changes from the baseline to adaptation. Therefore, participants were further divided into non-responders (i.e., $\leq$1% difference in all of the outcome variables after adaptation) or responders (>1% difference in at least one outcome variable after adaptation). The Shapiro–Wilks test (R package "rstatix") [41] was utilized to assess the distribution of each dependent variable (SLS SI%, step duration SI%, and stride length SI%), which indicated the gait variables from the sensors were non-normally distributed. Therefore, non-parametric tests were performed to compare gait asymmetry indices (SLS SI%, step duration SI%, and stride length SI%) at the baseline to adaptation within the groups and adaptation between groups, respectively. The Wilcoxon Signed-Rank and Mann–Whitney U tests (R package "stats") were performed for paired samples and independent samples indicated by test statistics "V" and "W", respectively. Only the 10th minute of each walking condition was used in the analyses. To assess the strength of the association between groups and gait asymmetry indices, we calculated the Wilcoxon Q effect size (R package "rstatix"). As the sample size is small due to the further division of groups based on if participants were categorized as a responder or not in addition to the data not having a normal distribution, the data were bootstrapped (5000 samples) with a bias correction for confidence intervals (CIs) to calculate effect sizes. The values for the Wilcoxon Q effect size are interpreted as follows: $r < 0.30$ = small effect, r = 0.30–0.49 = medium effect, and $r \geq 0.50$ = large effect [42]. For all analyses, the alpha level was set a priori at 0.05.

3. Results

A one-way ANOVA revealed a significant difference between the preferred walking speed between the groups ($p = 0.011$) (Table 1). A post hoc test with Bonferroni corrections showed the control group had a faster preferred walking speed than the small and large asymmetry-group responders ($p = 0.0075$ and $p = 0.0014$, respectively). No other statistically significant differences for the remaining demographic variables were observed.

The Wilcoxon Signed-Rank tests revealed, in the small asymmetry-group responders from the baseline to adaptation, that the single limb support SI% (V = 34, $p = 0.003$) and step duration SI% (V = 50, $p = 0.02$) increased, but stride length SI% was not significantly different (V = 88, $p = 0.273$) (Figure 2). Also, a large effect size was observed with the single limb support SI% (r = 0.59, CI [0.13, 0.8]); a medium effect size for step duration SI% (r = 0.46, CI [0.04, 0.74]); and a small effect size for stride length SI% (r = 0.14, CI [0, 0.43]). The Wilcoxon Signed-Rank tests revealed significant differences in the large asymmetry-group responders from the baseline to adaptation, as the single limb support SI% (V = 43, $p = 0.003$); step duration SI% (V = 41, $p = 0.002$); and stride length SI% (V = 33, $p = 0.0007$) increased in the adaptation condition. Large effect sizes were observed for SLS SI% (r = 0.58, CI [0.16, 0.79]); step duration SI% (r = 0.59, CI [0.11, 0.81]); and stride length SI% (r = 0.65, CI [0.29, 0.83]).

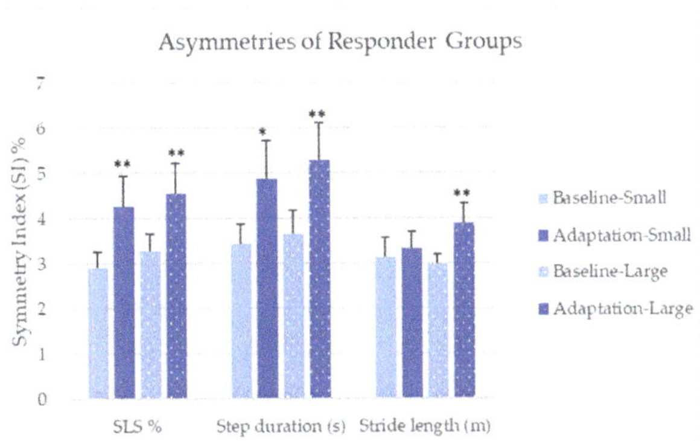

Figure 2. Comparison of baseline to adaptation in the small and large responder groups—means (SE). Note: SLS % = Single limb support %; * indicates $p < 0.05$; ** indicates $p < 0.01$.

The Wilcoxon Signed-Rank tests revealed, in the small asymmetry-group non-responders from the baseline to adaptation, no statistically significant differences in single limb support

SI% (V = 48, p = 0.765); step duration SI% (V = 36, p = 0.425); and stride length SI% (V = 39, p = 0.515) (Figure 3). Also, small effect sizes were observed with the single limb support SI% (r = 0.20, CI [0, 0.59]) and step duration SI% (r = 0.07, CI [0, 0.18]). The effect is too small for stride length SI% to calculate confidence intervals. The Wilcoxon Signed-Rank tests revealed no significant differences in the large asymmetry-group non-responders from the baseline to adaptation for single limb support SI% (V = 17, p = 0.161); step duration SI% (V = 28, p = 0.539); and stride length SI% (V = 18, p = 0.188). However, moderate effect sizes were observed for SLS SI% (r = 0.34, CI [0.02, 0.79]) and stride length SI% (r = 0.31, CI [0.02, 0.76]). The effect is too small for step duration SI% to calculate confidence intervals.

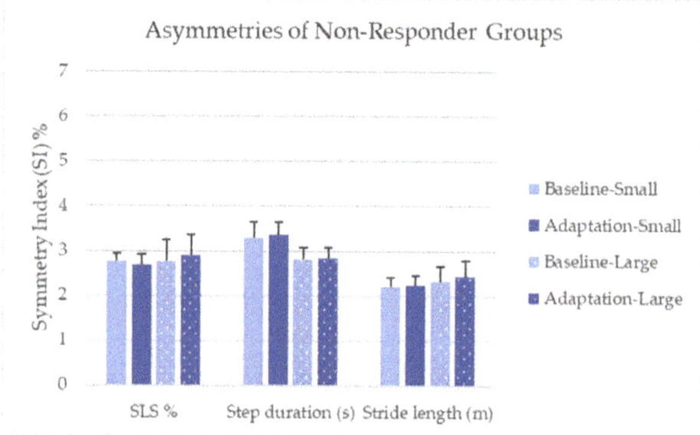

Figure 3. Comparison of baseline to adaptation in the small and large asymmetry non-responder groups—means (SE). Note: SLS % = Single limb support %.

The Wilcoxon Rank-Sum tests (i.e., Mann–Whitney U test) to compare the small and large asymmetry-group responders on the gait asymmetry metrics revealed no statistical differences between groups on SLS SI% (W = 193, p = 0.254, CI [$-\infty$, 0.480]); step duration SI% (W = 193, p = 0.122, CI [$-\infty$, 0.219]); and stride length SI% (W = 177, p = 0.144, CI [$-\infty$, 0.230)] (Figure 4). Based on our criteria for responders, three control participants were categorized as responders; however, when comparing the control responders to the non-responders, no significant differences were shown. Thus, the control group was recombined to include responders and non-responders. The control group did not significantly change from the baseline to adaptation in single limb support SI%, step duration SI%, or stride length SI% (Figure 5).

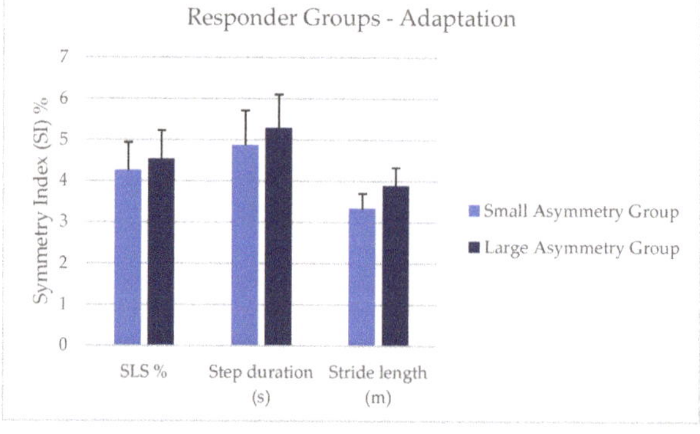

Figure 4. Comparison of small and large asymmetry-responder groups during the adaptation condition. Note: SLS % = Single limb support %.

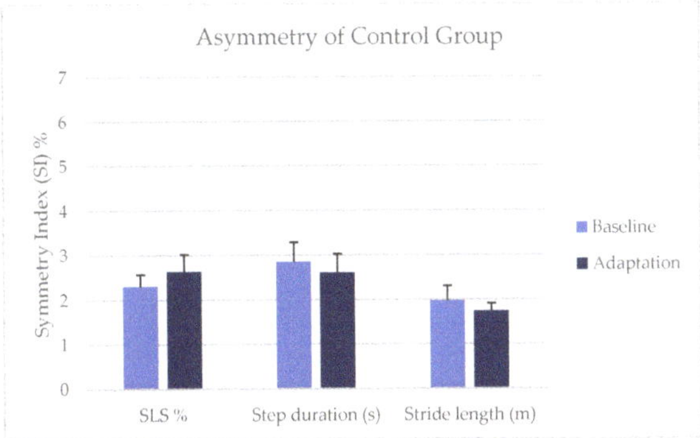

Figure 5. Comparison of asymmetries from baseline to adaptation in the control group. Note: SLS % = Single limb support %.

4. Discussion

The aim of this study was to examine the extent to which healthy adults can synchronize to an asymmetric visual cue during a 10-min treadmill walking session with a small or large gait asymmetry using inertial sensors. This study specifically examined the 10th minute of baseline and adaptation. We hypothesized that gait asymmetries would increase from the baseline to adaptation. Our hypothesis was mainly supported by the responder groups. In the small and large asymmetry-group responders, single limb support SI% and step duration SI% both increased during adaptation, indicating that mimicking an asymmetric cue can alter gait symmetry in healthy adults and that the inertial sensors are sensitive enough to detect small changes in asymmetry. The increases in single limb support asymmetry observed in the responder groups are indirectly supported by the literature, as changes were previously shown with double support asymmetry with split-belt treadmill training [16,18]. The changes in single limb support % and step duration asymmetry suggest temporal measures may be altered with an asymmetrical visual cue in healthy adults, similar to what Roemmich and colleagues [18] reported in their healthy young adult group. However, temporal changes may be harder to alter in clinical or aging populations [5,18,21].

Interestingly, the stride length SI% did not increase for the responders in the small asymmetry-responder group but did increase in the large asymmetry-responder group. The lack of changes to symmetry in the small asymmetry group may be due to the fact the small asymmetry visual cue was a subthreshold of what is clinically observable as having an asymmetric gait. The small asymmetry visual was a ratio of 1.4–1, while clinically observable asymmetry is a 1.5–1 ratio [2]. The asymmetrical visual cue ratio was based on speed, which may suggest participants were primarily focused with the timing of synchronizing to the visual cue rather than the spatial aspect such as in the stride length. Although, the difference between the speed of the visual cue and the participants' preferred walking speed may play a role. The large asymmetry ratio is similar to what is seen in the split-belt treadmill literature [20,43], so this may be why changes in stride length were seen only in the large asymmetry group and not the small asymmetry group.

In addition, we hypothesized that participants in the large asymmetry group would exhibit greater gait asymmetries than the small asymmetry group. Our hypothesis was not supported by our results, as no differences between the small and large asymmetry groups were found, even though we examined the responder groups. The findings were surprising, as the large asymmetrical visual cue had an obvious asymmetrical gait pattern, while the small asymmetrical visual cue was not as easily noticeable. In contrast to our findings, Reisman and colleagues [20] reported greater gait asymmetries for larger speed ratios of the belts. While we did not see differences when comparing the small and large responder groups, the magnitude of effect sizes varied between the small and large asymmetry groups.

The effect size of the single limb support SI% was a large effect size for the small and large asymmetry-responder groups. Step duration SI% was a moderate effect for the small asymmetry-responder group and large for the large asymmetry-responder group. The stride length SI% was small for the small asymmetry-responder group and large for the large asymmetry-responder group. The magnitude of the effect size indicates that the large asymmetrical visual cue has a greater impact on adaptation than the small asymmetrical visual cue. No significant differences were observed in the non-responder groups on gait asymmetries; although, the small asymmetry non-responder groups have small effect sizes, while the large asymmetry non-responder group showed medium effect sizes. As healthy adults have relatively symmetrical gait patterns yet are able to adapt to visual cues, whether it was the small or large asymmetrical ratio suggests that these novel methods were effective in changing gait patterns. The asymmetrically walking visual cues increased gait asymmetries in healthy adults, providing evidence that asymmetrically walking visual cues could be used to alter gait asymmetry in clinical populations without a split-belt treadmill.

Interestingly, 20 out of the 64 experimental group participants did not respond to the visual cue. Greater changes in gait are seen in early adaptation with split-belt treadmill training, so the lack of response may be because participants adapted early to the cue, since this study only examined the 10th minute of adaptation [18,20,44]. However, attention during the adaptation condition may play a role in whether a participant adapted to the visual cue or not. Although the health history questionnaire asked and excluded anyone with a neurological or cognitive disease or disorder, we did not explicitly ask nor exclude anyone with attentional deficit disorders. Attentional deficit disorders are very common and may not be thought of as a neurological disorder by this study's participants, as research on how attentional disorders in adults can affect how they walk is limited. However, research on children with attentional disorders show greater stride-to-stride variability during walking but reduced variability with dual tasks [45].

Motor-learning principles such as the rate of adaptation should also be considered in future research, as this study only focused on the 10th minute of adaptation. The rate of adaptation can tell us more information of the learning process when adapting to a novel gait task. The literature involving adaptation with split-belt treadmills shows immediate adaptation (i.e., the greatest differences) compared to the end of adaptation, as most healthy participants move back toward symmetry [20,44]. Thus, early adaptation should be compared to late adaptation to examine whether gait asymmetries were greater during the early adaptation condition. Also, Fitts' law should be considered as well. This study used novel methods, and speed may be an important component influencing the accuracy of mimicking the visual cue, which could contribute to the difficulty of the task [46]. Bootsma and colleagues [47] postulated that the magnitude of motor skill learning on task difficulty may be mediated by the perceived mental workload. Most gait adaptation studies have not asked participants how difficult they felt the task was, which may be impacting the accuracy or optimization of adaptation. A study that did examine perceived exertion reported that participants were divided into responders and non-responders, and the responders indicated a greater perceived exertion with overground walking than the non-responders on the first day of training [17].

A few limitations exist within our study. The sample size was small once the groups were further divided into responders and non-responders. We only examined the 10th minute of each condition, so early adaptation was not examined. Our participants were younger healthy adults, so our study lacks generalizability to a clinical population's ability to adapt to a visual cue.

5. Conclusions

Asymmetrically walking visual cues can be used to alter gait symmetry in healthy adults. Both a small and large gait asymmetry-ratio visual cue can increase gait asymmetry. Inertial sensors are sensitive to detecting small changes in gait asymmetries and allow for greater clinical applicability. Future research should investigate why some people did not respond to the visual cue and examine if preferred walking speed, difficulty, or mental workload were factors in the lack of a response for some participants. Also, future research should investigate how visual cueing can be utilized to reduce gait asymmetries in clinical populations and use inertial sensors to detect gait changes in and out of the lab environment.

Supplementary Materials: The following supporting information can be downloaded at: https://www.mdpi.com/article/10.3390/biomechanics4020024/s1, Video S1: 1.4 ratio, Video S2: 1.9 ratio.

Author Contributions: Conceptualization, K.G.C. and C.K.R.; methodology, K.G.C., L.D.R., S.E.R. and C.K.R.; software, K.G.C.; validation, K.G.C. and C.K.R.; formal analysis, K.G.C.; investigation, K.G.C.; resources, C.K.R.; data curation, K.G.C.; writing—original draft preparation, K.G.C.; writing—review and editing, K.G.C., L.D.R., S.E.R. and C.K.R.; visualization, K.G.C.; supervision, C.K.R.; project administration, K.G.C. All authors have read and agreed to the published version of the manuscript.

Funding: This research received no external funding.

Institutional Review Board Statement: This study was conducted in accordance with the Declaration of Helsinki and approved by the Institutional Review Board of the University of North Carolina Greensboro (IRB-FY23-72, approval date: 18 October 2022).

Informed Consent Statement: Informed consent was obtained from all subjects involved in this study.

Data Availability Statement: Data will be made available upon reasonable request to the corresponding author.

Acknowledgments: The authors would like to thank Michael D. Lewek for providing the data that were used to create the visual cues and all who volunteered to participate in this study.

Conflicts of Interest: The authors declare no conflicts of interest.

References

1. Jørgensen, L.; Crabtree, N.J.; Reeve, J.; Jacobsen, B.K. Ambulatory Level and Asymmetrical Weight Bearing after Stroke Affects Bone Loss in the Upper and Lower Part of the Femoral Neck Differently: Bone Adaptation after Decreased Mechanical Loading. *Bone* **2000**, *27*, 701–707. [CrossRef]
2. Patterson, K.K.; Parafianowicz, I.; Danells, C.J.; Closson, V.; Verrier, M.C.; Staines, W.R.; Black, S.E.; McIlroy, W.E. Gait Asymmetry in Community-Ambulating Stroke Survivors. *Arch. Phys. Med. Rehabil.* **2008**, *89*, 304–310. [CrossRef]
3. Yogev, G.; Plotnik, M.; Peretz, C.; Giladi, N.; Hausdorff, J.M. Gait Asymmetry in Patients with Parkinson's Disease and Elderly Fallers: When Does the Bilateral Coordination of Gait Require Attention? *Exp. Brain Res.* **2007**, *177*, 336–346. [CrossRef]
4. Awad, L.N.; Palmer, J.A.; Pohlig, R.T.; Binder-Macleod, S.A.; Reisman, D.S. Walking Speed and Step Length Asymmetry Modify the Energy Cost of Walking After Stroke. *Neurorehabilit. Neural Repair* **2015**, *29*, 416–423. [CrossRef]
5. Lewek, M.D.; Bradley, C.E.; Wutzke, C.J.; Zinder, S.M. The Relationship Between Spatiotemporal Gait Asymmetry and Balance in Individuals With Chronic Stroke. *J. Appl. Biomech.* **2014**, *30*, 31–36. [CrossRef]
6. Wei, T.-S.; Liu, P.-T.; Chang, L.-W.; Liu, S.-Y. Gait Asymmetry, Ankle Spasticity, and Depression as Independent Predictors of Falls in Ambulatory Stroke Patients. *PLoS ONE* **2017**, *12*, e0177136. [CrossRef]
7. Alexander, B.H.; Rivara, F.P.; Wolf, M.E. The Cost and Frequency of Hospitalization for Fall-Related Injuries in Older Adults. *Am. J. Public Health* **1992**, *82*, 1020–1023. [CrossRef]
8. Axer, H.; Axer, M.; Sauer, H.; Witte, O.W.; Hagemann, G. Falls and Gait Disorders in Geriatric Neurology. *Clin. Neurol. Neurosurg.* **2010**, *112*, 265–274. [CrossRef]
9. Contreras, A.; Grandas, F. Risk of Falls in Parkinson's Disease: A Cross-Sectional Study of 160 Patients. *Park. Dis.* **2012**, *2012*, 362572. [CrossRef]
10. Mahlknecht, P.; Kiechl, S.; Bloem, B.R.; Willeit, J.; Scherfler, C.; Gasperi, A.; Rungger, G.; Poewe, W.; Seppi, K. Prevalence and Burden of Gait Disorders in Elderly Men and Women Aged 60–97 Years: A Population-Based Study. *PLoS ONE* **2013**, *8*, e69627. [CrossRef]
11. Malone, L.A.; Bastian, A.J. Spatial and Temporal Asymmetries in Gait Predict Split-Belt Adaptation Behavior in Stroke. *Neurorehabilit. Neural Repair* **2014**, *28*, 230–240. [CrossRef]
12. Sterling, D.A.; O'Connor, J.A.; Bonadies, J. Geriatric Falls: Injury Severity Is High and Disproportionate to Mechanism. *J. Trauma Acute Care Surg.* **2001**, *50*, 116–119. [CrossRef]
13. Wagner, L.M.; Phillips, V.L.; Hunsaker, A.E.; Forducey, P.G. Falls among Community-Residing Stroke Survivors Following Inpatient Rehabilitation: A Descriptive Analysis of Longitudinal Data. *BMC Geriatr.* **2009**, *9*, 46. [CrossRef]

14. Brandstater, M.E.; de Bruin, H.; Gowland, C.; Clark, B.M. Hemiplegic Gait: Analysis of Temporal Variables. *Arch. Phys. Med Rehabil.* **1983**, *64*, 583–587.
15. Nanhoe-Mahabier, W.; Snijders, A.H.; Delval, A.; Weerdesteyn, V.; Duysens, J.; Overeem, S.; Bloem, B.R. Split-Belt Locomotion in Parkinson's Disease with and without Freezing of Gait. *Neuroscience* **2013**, *236*, 110–116. [CrossRef]
16. Reisman, D.S.; Wityk, R.; Silver, K.; Bastian, A.J. Locomotor Adaptation on a Split-Belt Treadmill Can Improve Walking Symmetry Post-Stroke. *Brain* **2007**, *130*, 1861–1872. [CrossRef]
17. Reisman, D.S.; McLean, H.; Keller, J.; Danks, K.A.; Bastian, A.J. Repeated Split-BeltTreadmill Training Improves Poststroke Step Length Asymmetry. *Neurorehabilit. Neural Repair* **2013**, *27*, 460–468. [CrossRef]
18. Roemmich, R.T.; Nocera, J.R.; Stegemöller, E.L.; Hassan, A.; Okun, M.S.; Hass, C.J. Locomotor Adaptation and Locomotor Adaptive Learning in Parkinson's Disease and Normal Aging. *Clin. Neurophysiol.* **2014**, *125*, 313–319. [CrossRef]
19. Meder, K.G.; LoJacono, C.T.; Rhea, C.K. A Systematic Review of Non-Pharmacological Interventions to Improve Gait Asymmetries in Neurological Populations. *Symmetry* **2022**, *14*, 281. [CrossRef]
20. Reisman, D.S.; Block, H.J.; Bastian, A.J. Interlimb Coordination During Locomotion: What Can Be Adapted and Stored? *J. Neurophysiol.* **2005**, *94*, 2403–2415. [CrossRef]
21. Lewek, M.D.; Braun, C.H.; Wutzke, C.; Giuliani, C. The Role of Movement Errors in Modifying Spatiotemporal Gait Asymmetry Post Stroke: A Randomized Controlled Trial. *Clin. Rehabil.* **2018**, *32*, 161–172. [CrossRef]
22. Maidan, I.; Patashov, D.; Shustak, S.; Fahoum, F.; Gazit, E.; Shapiro, B.; Levy, A.; Sosnik, R.; Giladi, N.; Hausdorff, J.M.; et al. A New Approach to Quantifying the EEG during Walking: Initial Evidence of Gait Related Potentials and Their Changes with Aging and Dual Tasking. *Exp. Gerontol.* **2019**, *126*, 110709. [CrossRef]
23. Pozzi, N.G.; Canessa, A.; Palmisano, C.; Brumberg, J.; Steigerwald, F.; Reich, M.M.; Minafra, B.; Pacchetti, C.; Pezzoli, G Volkmann, J.; et al. Freezing of Gait in Parkinson's Disease Reflects a Sudden Derangement of Locomotor Network Dynamics. *Brain* **2019**, *142*, 2037–2050. [CrossRef]
24. Hasegawa, N.; Shah, V.V.; Carlson-Kuhta, P.; Nutt, J.G.; Horak, F.B.; Mancini, M. How to Select Balance Measures Sensitive to Parkinson's Disease from Body-Worn Inertial Sensors-Separating the Trees from the Forest. *Sensors* **2019**, *19*, 3320. [CrossRef]
25. Cano Porras, D.; Siemonsma, P.; Inzelberg, R.; Zeilig, G.; Plotnik, M. Advantages of Virtual Reality in the Rehabilitation of Balance and Gait: Systematic Review. *Neurology* **2018**, *90*, 1017–1025. [CrossRef]
26. Janeh, O.; Fründt, O.; Schönwald, B.; Gulberti, A.; Buhmann, C.; Gerloff, C.; Steinicke, F.; Pötter-Nerger, M. Gait Training in Virtual Reality: Short-Term Effects of Different Virtual Manipulation Techniques in Parkinson's Disease. *Cells* **2019**, *8*, 419. [CrossRef]
27. Keshner, E.A.; Lamontagne, A. The Untapped Potential of Virtual Reality in Rehabilitation of Balance and Gait in Neurological Disorders. *Front. Virtual Real.* **2021**, *2*, 6. [CrossRef]
28. Liu, L.Y.; Sangani, S.; Patterson, K.K.; Fung, J.; Lamontagne, A. Real-Time Avatar-Based Feedback to Enhance the Symmetry of Spatiotemporal Parameters After Stroke: Instantaneous Effects of Different Avatar Views. *IEEE Trans. Neural Syst. Rehabil. Eng.* **2020**, *28*, 878–887. [CrossRef]
29. Meerhoff, L.A.; De Poel, H.J.; Jowett, T.W.D.; Button, C. Walking with Avatars: Gait-Related Visual Information for Following a Virtual Leader. *Hum. Mov. Sci.* **2019**, *66*, 173–185. [CrossRef]
30. Rhea, C.K.; Kiefer, A.W.; D'Andrea, S.E.; Warren, W.H.; Aaron, R.K. Entrainment to a Real Time Fractal Visual Stimulus Modulates Fractal Gait Dynamics. *Hum. Mov. Sci.* **2014**, *36*, 20–34. [CrossRef]
31. Soczawa-Stronczyk, A.A.; Bocian, M. Gait Coordination in Overground Walking with a Virtual Reality Avatar. *R. Soc. Open Sci.* **2020**, *7*, 200622. [CrossRef] [PubMed]
32. Felsberg, D.T.; Rhea, C.K. Spontaneous Interpersonal Synchronization of Gait: A Systematic Review. *Arch. Rehabil. Res. Clin. Transl.* **2021**, *3*, 100097. [CrossRef] [PubMed]
33. Kaufman, K.R.; Miller, L.S.; Sutherland, D.H. Gait Asymmetry in Patients with Limb-Length Inequality. *J. Pediatr. Orthop.* **1996**, *16*, 144–150. [CrossRef] [PubMed]
34. Khamis, S.; Carmeli, E. Relationship and Significance of Gait Deviations Associated with Limb Length Discrepancy: A Systematic Review. *Gait Posture* **2017**, *57*, 115–123. [CrossRef] [PubMed]
35. Dingwell, J.B.; Marin, L.C. Kinematic Variability and Local Dynamic Stability of Upper Body Motions When Walking at Different Speeds. *J. Biomech.* **2006**, *39*, 444–452. [CrossRef] [PubMed]
36. Reisman, D.S.; Wityk, R.; Silver, K.; Bastian, A.J. Split-Belt Treadmill Adaptation Transfers to Overground Walking in Persons Poststroke. *Neurorehabilit. Neural Repair* **2009**, *23*, 735–744. [CrossRef] [PubMed]
37. Morris, R.; Stuart, S.; McBarron, G.; Fino, P.C.; Mancini, M.; Curtze, C. Validity of MobilityLab (Version 2) for Gait Assessment in Young Adults, Older Adults and Parkinson's Disease. *Physiol. Meas.* **2019**, *40*, 095003. [CrossRef] [PubMed]
38. Faul, F.; Erdfelder, E.; Lang, A.-G.; Buchner, A. G*Power 3: A Flexible Statistical Power Analysis Program for the Social, Behavioral and Biomedical Sciences. *Behav. Res. Methods* **2007**, *39*, 175–191. [CrossRef] [PubMed]
39. R Core Team. *R: A Language and Environment for Statistical Computing*; R Foundation for Statistical Computing: Vienna, Austria, 2023.
40. Błażkiewicz, M.; Wiszomirska, I.; Wit, A. Comparison of Four Methods of Calculating the Symmetry of Spatial-Temporal Parameters of Gait. *Acta Bioeng. Biomech.* **2014**, *16*, 29–35. [CrossRef]
41. Kassambara, A. rstatix: Pipe-Friendly Framework for Basic Statistical Tests. R Package Version 0.7.2, 2023. Available online: https://cran.r-project.org/web/packages/rstatix/index.html (accessed on 10 May 2024).

42. Wilcox, R. A Robust Nonparametric Measure of Effect Size Based on an Analog of Cohen's d, Plus Inferences About the Median of the Typical Difference. *J. Mod. Appl. Stat. Methods* **2019**, *17*, eP2726. [CrossRef]
43. Prokop, T.; Berger, W.; Zijlstra, W.; Dietz, V. Adaptational and Learning Processes during Human Split-Belt Locomotion: Interaction between Central Mechanisms and Afferent Input. *Exp. Brain Res.* **1995**, *106*, 449–456. [CrossRef] [PubMed]
44. Bruijn, S.M.; Van Impe, A.; Duysens, J.; Swinnen, S.P. Split-Belt Walking: Adaptation Differences between Young and Older Adults. *J. Neurophysiol.* **2012**, *108*, 1149–1157. [CrossRef] [PubMed]
45. Leitner, Y.; Barak, R.; Giladi, N.; Peretz, C.; Eshel, R.; Gruendlinger, L.; Hausdorff, J.M. Gait in Attention Deficit Hyperactivity Disorder. *J. Neurol.* **2007**, *254*, 1330–1338. [CrossRef]
46. Schmidt, R.A.; Lee, T.D.; Winstein, C.J.; Wulf, G.; Zelaznik, H.N. *Motor Control and Learning: A Behavioral Emphasis*; Human Kinetics: Champaign, IL, USA, 2019; ISBN 978-1-4925-4775-4.
47. Bootsma, J.M.; Hortobágyi, T.; Rothwell, J.C.; Caljouw, S.R. The Role of Task Difficulty in Learning a Visuomotor Skill. *Med. Sci. Sports Exerc.* **2018**, *50*, 1842. [CrossRef] [PubMed]

biomechanics

MDPI

Article

Concurrent Validity of Depth-Sensor-Based Quantification of Compensatory Movements during the Swing Phase of Gait in Healthy Individuals

Kento Kusuda [1], Shigehito Matsubara [1,2], Daisuke Noguchi [3], Moe Kuwahara [3], Hiroomi Hamasaki [3], Toshihiro Miwa [3], Toru Maeda [3], Toshihito Nakanishi [4], Shogo Ninomiya [5] and Keita Honda [1,2,*]

1 Graduate School of Health Sciences, Kumamoto Health Science University, 325 Izumi-machi, Kita-ku, Kumamoto 861-5598, Japan; happylifekusuda@gmail.com (K.K.)
2 Department of Rehabilitation, Kumamoto Health Science University, 325 Izumi-machi, Kita-ku, Kumamoto 861-5598, Japan
3 Department of Physical Therapy, Kumamoto Kinoh Hospital, 6-8-1 Yamamuro, Kita-ku, Kumamoto 860-8518, Japan
4 Department of Rehabilitation, Kumamoto Kinoh Hospital, 6-8-1 Yamamuro, Kita-ku, Kumamoto 860-8518, Japan
5 Department of Physical Therapy, School of Health Science, Tokyo International University, 1-13-1 Matobakita, Kawagoe-shi, Saitama 350-1197, Japan
* Correspondence: honda.kt@kumamoto-hsu.ac.jp; Tel.: +81-96-275-2162

Abstract: The advancement in depth-sensor technology increased the potential for the clinical use of markerless three-dimensional motion analysis (3DMA); however, the accurate quantification of depth-sensor-based 3DMA on gait characteristics deviating from normal patterns is unclear. This study investigated the concurrent validity of the measurements of compensatory movements measured by depth-sensor-based 3DMA compared to those measured by marker-based 3DMA. We induced swing-phase compensatory movements due to insufficient toe clearance by restricting unilateral ankle and knee joint movements in healthy individuals. Thirty-two healthy young adults (nineteen males, aged 20.4 ± 2.0 years, height 164.4 ± 9.8 cm, weight 60.0 ± 9.3 kg [average ± standard deviation]) walked the 6 m walkway in slow speed, very slow speed, and knee–ankle–foot orthosis (KAFO; participants wore KAFOs on the right leg) conditions. Gait kinematics were measured with marker-based and depth-sensor-based 3DMA systems. The intraclass correlation coefficient ($ICC_{3,1}$) was used to measure the relative agreement between depth-sensor-based and marker-based 3DMA and demonstrated good or moderate validity for swing-phase compensatory movement measurement. Additionally, the $ICC_{2,1}$ measured absolute agreement between the systems and showed lower validity than the $ICC_{3,1}$. The measurement errors for contralateral vaulting, trunk lateral flexion, hip hiking, swing-side hip abduction, and circumduction between instruments were 0.01 m, 1.30°, 1.99°, 2.37°, and 1.53°, respectively. Depth-sensor-based 3DMA is useful for determining swing-phase compensatory movements, although the possibility of missing a slight measurement error of 1–2° must be considered.

Keywords: markerless motion capture; depth sensor; validity study; abnormal gait pattern; kinematic

1. Introduction

The decreased walking speed in post-stroke patients is attributed to functional impairments, including motor paralysis, muscle weakness, spasticity, and sensory disturbance. Stroke-induced decreased walking speed is a serious problem in daily life, causing reduced quality of life and limited social participation [1–4]. Perry et al. revealed that the walking speeds of household-, limited community-, and community-walker post-stroke patients are <0.4, 0.4–0.8, and >0.8 m/s, respectively [5]. Similarly, Fulk et al. [6] indicated that the cutoff value for discriminating between household and community walkers is a comfortable walking speed of 0.49 m/s. Additionally, 43.08% (190/441) of post-stroke patients are household walkers, indicating that a large number of these patients are unable to walk independently outside the home [6]. Therefore, improving walking speed is crucial in post-stroke gait rehabilitation.

Gait training has helped improve walking speed. Treadmill gait training for community-dwelling post-stroke patients has improved walking speed by increasing ankle plantar flexion moment during the paretic stance phase [7]. Conversely, Patterson et al. [8] revealed no improvement in step-length asymmetry and swing-phase time asymmetry between the paretic and nonparetic sides despite walking speed improvement by gait training. This result indicates that walking speed improvement is attributed not only to the enhancement of functional impairment on the paretic side but also to an increased compensatory movement by the nonparetic side [8]. Spatiotemporal parameter asymmetry during gait was associated with increased energy costs during gait [9] and reduced bone density in the femoral neck of the paretic side [10]. Therefore, the gait pattern assessment findings should be considered when providing gait rehabilitation to post-stroke patients to improve walking speed and prevent secondary disability.

Typical gait measures include lower limb circumduction, hip hiking, and trailing limb angle [11–13]. Given that these measures are calculated from the relative distance of each joint position and angle of each segment, quantifying the movement of body landmarks during gait is necessary. The gold standard method for quantifying the movement of body landmarks is marker-based optical three-dimensional motion analysis (3DMA), which uses multiple infrared cameras to measure the 3D coordinates of reflective markers attached to the body landmarks [11–13]. This method accurately measures the 3D coordinates of the landmarks; thus, marker-based 3DMA evaluates detailed gait characteristics [11–13]. However, the need for expensive measurement equipment and the burden of attaching numerous reflective markers to the body during measurements limit its use in clinical practice [14]. Advancement in depth-sensor technology for quantitative gait pattern assessment can solve the aforementioned problem in clinical practice [15,16]. Azure Kinect, a typical markerless 3DMA system based on depth-sensor technology, captures the contour and surface irregularities of a subject using a depth sensor and uses machine learning-based skeletal recognition technology to process this depth image. This process allows the estimation of the 3D coordinates of body landmarks [17]. Thus, Azure Kinect is advantageous for clinical implementation due to the elimination of the need to attach reflective markers, although it may provide less accurate measurements compared to the marker-based 3DMA system.

The concurrent validity of depth-sensor-based 3DMA, measured against marker-based 3DMA, for measuring spatiotemporal and kinematic parameters during steady-state gait in healthy individuals has been previously investigated [18,19]. A good concurrent validity in the depth-sensor-based assessment of spatiotemporal variables and each lower limb joint angle in the sagittal plane, except for the ankle joint angle, was reported [19]. Conversely, the accuracy of measurement of the deviation of abnormal gait patterns from normal patterns using depth-sensor-based 3DMA is unclear. Decreased knee flexion and ankle dorsiflexion during the paretic swing phase reduce toe clearance in post-stroke patients with hemiparesis, which induces compensatory movements during the swing phase, such as hip hiking and circumduction [20]. Although previous studies have assessed the compensatory movements during the swing phase by marker-based 3DMA (Table 1), it is unclear whether these parameters can be assessed by depth-sensor-based 3DMA. These abnormal gait patterns were observed in healthy individuals during gait with knee joint motion restricted by the use of an immobilizing orthosis. Zissimopoulos et al. [21] have demonstrated that pelvic hiking in healthy individuals with knee braces was significantly larger than that of individuals without knee braces. Akbus et al. [22] have reported that the pelvic hiking during the swing phase in healthy individuals wearing a knee–ankle–foot orthosis (KAFO) was not different from that of post-stroke patients. Thus, we induced an abnormal gait pattern in healthy individuals by directing them to wear a KAFO to their unilateral lower limbs and verified the concurrent validity of the depth-sensor-based assessment.

Table 1. Methods for assessing compensatory movements during the swing phase.

Study	Participants	Indicators of Compensatory Movements during Swing Phase	Equipment
Kerrigan et al. [11]	Post-stroke patients (n = 23), healthy people (n = 23)	Peak values during paretic swing phase (coronal pelvic angle, transverse pelvic rotation, hip adduction, hip abduction, thigh adduction, and thigh abduction), values at mid-swing (coronal pelvic angle, transverse pelvic angle, bilateral coronal hip angle, and affected thigh angle)	Marker-based 3DMA
Stanhope et al. [12]	Post-stroke patients (n = 21)	Peak paretic pelvic tilt angle, peak paretic hip abduction angle, toe displacement	Marker-based 3DMA
Tyrell et al. [13]	Post-stroke patients (n = 22)	Hip hiking (the angle in the frontal plane between the pelvis position in static standing and the maximum deviation from that position in the swing phase), circumduction (maximum lateral difference between the position of the heel marker in the stance phase and the same heel marker position in the swing phase immediately following the stance phase)	Marker-based 3DMA

The present study aimed to investigate the concurrent validity of depth-sensor-based 3DMA for quantifying abnormal gait patterns during the swing phase as compared to a reference marker-based 3DMA among healthy individuals. We hypothesized that depth-sensor-based 3DMA would be effective in quantifying compensatory movements due to insufficient toe clearance. Verifying that depth-sensor-based 3DMA is able to quantify abnormal gait patterns during the swing phase with the equivalent accuracy as the marker-based 3DMA would result in the development of rehabilitation practices based on objective gait assessment.

2. Materials and Methods

2.1. Participants

The present study included 32 healthy young adults (Table 2) and excluded those with musculoskeletal pain, previous lower extremity or trunk surgery, or neurological disease history that could affect motor control.

Table 2. Participant information.

Sex (male, female) [a]	32 (19/13)
Age (years) [b]	20.4 ± 2.0
Height (cm) [b]	164.4 ± 9.8
Weight (kg) [b]	60.0 ± 9.3

[a] Number of participants; [b] mean ± standard deviation.

2.2. Experimental Procedure

Participants were instructed to walk a 6 m walkway five times in each of the slow speed, very slow speed, and KAFO conditions. The cadence and step length were set at 90 bpm and 350 mm, respectively, in the slow condition and 70 bpm and 250 mm, respectively, in the very slow speed condition. We referenced the results of the spatiotemporal variables for post-stroke patients to determine the cadence and step length for each condition [23]. Post-stroke patients walking at speeds ranging from 0.5 to 1.4 km/h

(0.14–0.39 m/s) exhibited a mean cadence and paretic step length of 72.72 steps/min and 24.64 cm, respectively. Conversely, those walking at speeds between 1.5 and 2.4 km/h (0.42–0.67 m/s) demonstrated a mean cadence and paretic step length of 91.71 steps/min and 35.72 cm, respectively. Participants in the KAFO condition wore a KAFO (Modular Leg Brace NEO, TOKUDA Prosthetics and Orthotics Mfg Ltd., Kumamoto, Japan) on the right lower limb in addition to performing the tasks in the very slow speed condition. The KAFO was set with the knee and ankle joints in 0° extension and 0° dorsiflexion, respectively. Participants in the cadence control were instructed to step according to an electronic metronome set for each condition. The lengths of the first and second steps at the beginning of gait in the control of step length were indicated by lines tape-marked on the floor, and participants were instructed to maintain this step length. Data collections were conducted after practice using the abovementioned settings before each condition to allow participants to familiarize themselves with the task. After visually confirming that the participant was stepping onto the electronic metronome, the examiner verbally asked the participant if they were familiar with the task.

2.3. Data Collection

An eight-infrared-camera marker-based 3DMA system (MAC 3D, Motion Analysis Corporation, Santa Rosa, CA, USA) and a depth-sensor-based 3DMA system (Azure Kinect, Microsoft Corporation, Redmond, WA, USA) (ICpro-AK, Hu-tech Co., Ltd., Tama, Tokyo, Japan) were used to simultaneously measure the 3D coordinates of the body landmarks during gait (sampling frequencies of 100 and 30 Hz, respectively). The infrared cameras were placed 3 to 5 m away from the walkway. In this setup, a 500 mm calibration wand could be used for measurement, with a measurement error of <1 mm. The depth sensor was placed 7 m from the starting position of walking, i.e., 1 m behind the endpoint (Figure 1). The depth sensor was positioned in front of the participant, and the height from the floor to the sensor was set at 0.7 m. Prior to the actual experiment, the skeletal recognition of the depth sensor was not displayed when a 19 mm diameter reflective marker was attached to the participant's body, and the marker- and depth-sensor-based 3DMA systems were used in the measurement. Contrarily, the use of 9.5 mm diameter reflective markers significantly reduced the skeletal recognition error; thus, we used 9.5 mm diameter markers in the actual experiment. Prior to the actual experiments, the 3DMA skeletal recognition using the depth sensor started at a distance of approximately 5 m from the sensor. Given that artifacts were observed immediately after skeletal recognition, the analysis section was set from 1.8 to 3.5 m, similar to that used by Ferraris et al. [24]. A depth sensor, set up according to the instructions for the machine learning-based skeletal recognition technology produced by the Microsoft Corporation, estimated and collected 19 body landmark positions during gait [25] (Figure 2). In the marker-based 3DMA, 19 reflective markers were attached to the body landmarks to calculate the trunk and lower limb kinematics (Figure 3). An experienced physical therapist (K.H.) attached the reflective markers to the body landmarks.

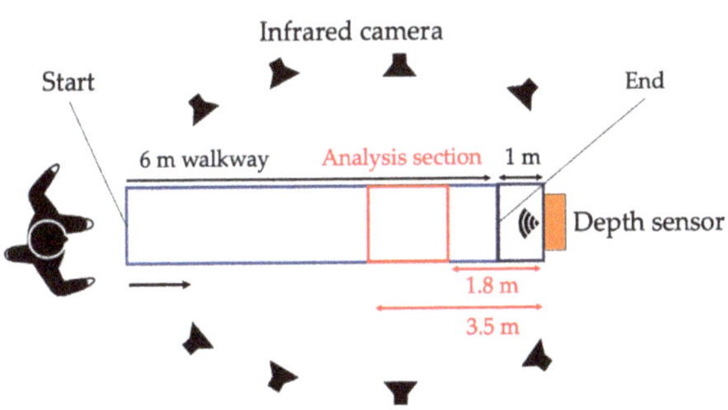

Figure 1. Overview of the experimental setup.

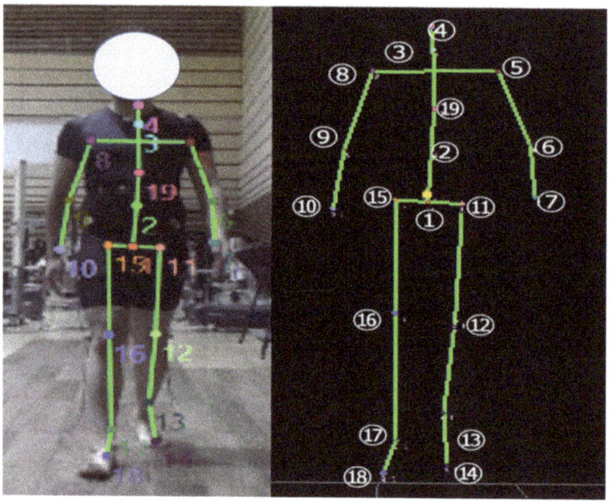

1. Pelvis
2. Spine naval
3. Neck
4. Head
5. Shoulder left
6. Elbow left
7. Wrist left
8. Shoulder right
9. Elbow right
10. Wrist right
11. Hip left
12. Knee left
13. Ankle left
14. Foot left
15. Hip right
16. Knee right
17. Ankle right
18. Foot right
19. Spine chest

Figure 2. Body landmarks derived from the skeletal recognition technology of Azure Kinect.

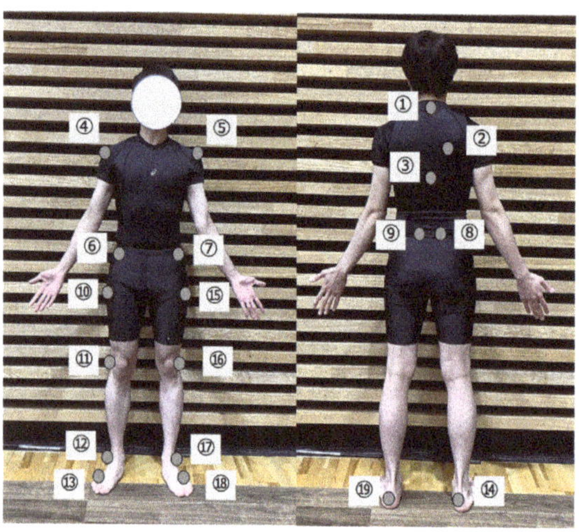

1. Spinous process of the 7th cervical vertebrae
2. The position in the middle of the right scapula
3. Spinous process of the 10th thoracic vertebrae
4. Right acromion
5. Left acromion
6. Right anterior superior iliac spines
7. Left anterior superior iliac spines
8. Right posterior superior iliac spines
9. Left posterior superior iliac spines
10. Right greater trochanters
11. Right lateral epicondyle of the knee
12. Right lateral malleoli
13. Right fifth metatarsal head of the foot
14. Right heel
15. Left greater trochanters
16. Left lateral epicondyle of the knee
17. Left lateral malleoli
18. Left fifth metatarsal head of the foot
19. Left heel

Figure 3. Marker locations for marker-based 3D motion analysis.

2.4. Failed Trial of Gait Analysis Using Depth-Sensor-Based 3DMA

Of the 32 participants, 10 were unable to walk under the KAFO condition because they could not correctly wear the KAFO, which inhibited them from walking even a single step. Hence, only 22 participants performed the gait trial in the KAFO condition, and 32 participants underwent the gait trial in the slow and very slow conditions. Failed trials (i.e., unsuccessful trials) were observed in 9 (28%), 11 (34%), and 5 (23%) participants in the slow, very slow, and KAFO conditions, respectively. The failed trials were defined as those in which the coordinates of the body landmarks could not be measured by depth-sensor-based 3DMA. After the measurement, the researchers (K.K. and K.H.) exported the coordinate data to determine whether the trial was a failure or not.

2.5. Data Reduction and Analysis

The analysis used three or more successful measurement trials out of five trials in each condition. A successful measurement trial was defined as a successful estimation of the 3D coordinate values of 19 points by the depth-sensor-based 3DMA. MATLAB R2019b (The Math Works, Inc., Natick, MA, USA) was used to calculate the subsequent data analysis process.

The 3D coordinate data measured by the depth sensor were resampled from 30 to 100 Hz using cubic spline interpolation to align the sampling frequencies between pieces of equipment. A zero-lag fourth-order Butterworth low-pass filter (cutoff frequency: 6 Hz) was used to smooth the 3D coordinate data from both pieces of equipment.

The landmarks of the lower limbs (i.e., hip, knee, ankle, and toe) in marker-based 3DMA were defined from the 3D coordinates of the measured reflective markers according to previous studies [26–28]. We analyzed the joint angles by projecting the 3D coordinates onto a two-dimensional plane. The trunk flexion–extension and lateral flexion angles in marker- and depth-sensor-based 3DMAs were identified as the angle of the longitudinal axis of the trunk in the sagittal and coronal planes in the global coordinate system. Additionally, the trunk rotation angle was determined as the angle of the mediolateral axis of the trunk in the transversal plane in the global coordinate system. The flexion–extension (dorsiflexion–plantar flexion) angles of the hip, knee, and ankle joints were defined as the angle between the longitudinal axes of the two segments adjacent to the reference joint in the sagittal plane. The leg extension angle was calculated as the angle of the line connecting the hip and ankle joint points in the global coordinate system in the sagittal plane [29].

The center of mass (CoM) was calculated using the 3D coordinate positions of the shanks, thighs, and trunk measured by each piece of equipment and body segment inertia parameters described elsewhere [30]. Foot information was removed from the CoM calculation because toe tracking on the depth sensor is generally very poor [19,31,32]. The CoM velocity was obtained by applying a three-point differential equation to time series data of CoM positions [30].

We analyzed the gait cycle of the right side, from the right initial contact (IC) to the next ipsilateral IC. The IC in the right gait cycle was identified as the point where the right ankle joint point is farthest forward from the pelvic center point [33,34]. The pelvic center point in the marker-based 3DMA was defined as the midpoint between the left and right posterior superior iliac spines. The toe-off (TO) was the point where the right ankle joint point is farthest backward from the pelvic center point [33,34].

2.6. Spatiotemporal Variables

The spatiotemporal variables calculated based on previous reports were as follows [18,19]: step width, right and left step lengths, stride length, cadence, gait velocity, stance phase time, swing phase time, gait cycle time, %stance phase, %swing phase, step length ratio, and swing time ratio.

2.7. *Trunk and Lower Limb Kinematics*

The mean and peak-to-peak values of the trunk angle in each plane in the trunk kinematics were calculated to determine the upper body sway and average posture during a gait cycle. The maximum (i.e., maximum extension or plantar flexion), minimum (i.e., maximum flexion or dorsiflexion), and peak-to-peak values of each joint angle in the sagittal plane in lower limb kinematics were calculated during a gait cycle. Additionally, the maximum value of the leg extension angle during a gait cycle was calculated as an indicator of propulsion [35].

2.8. *Whole-Body Kinematics*

The peak-to-peak value of the CoM position in the mediolateral and vertical axes was calculated to determine the whole-body sway during a gait cycle. Additionally, the margin of stability (MoS) was computed as the distance between the extrapolated CoM (XCoM) positions and ankle joint (i.e., boundaries of the base of support) to determine the dynamic stability during gait [36].

2.9. *Definition of Abnormal Gait Patterns*

The following five indices were calculated as abnormal gait patterns compensating for toe clearance during the swing phase: contralateral vaulting, trunk lateral flexion to the stance side, hip hiking, swing-side hip abduction, and circumduction. Contralateral vaulting, which is the elevation due to the lower limb on the stance side, was determined by subtracting the maximum height of the stance-side hip during 25–75% of the swing phase by the time-averaged height of the ipsilateral hip during the pre-swing phase (Figure 4a) [37]. The trunk lateral flexion to the stance side was calculated by subtracting the maximum trunk lateral flexion angle during 25–75% of the swing phase by the same time-averaged angle during the pre-swing phase (Figure 4b) [37]. Hip hiking is a pelvic raising movement on the swing limb due to stance-side hip abduction, which was determined by subtracting the maximum pelvic angle (i.e., the line connecting the left–right hip joint) in the coronal plane during 25–75% of the swing phase by the same time-averaged angle during the pre-swing phase (Figure 4c) [11]. The swing-side hip abduction was calculated by subtracting the maximum hip angle (i.e., angle of the thigh relative to the pelvis) in the coronal plane during 25–75% of the swing phase by the same time-averaged angle during the pre-swing phase (Figure 4d) [11,12]. Circumduction is the extent to which the lower extremity deviates from its normal trajectory in the coronal plane as a result of swing-side hip abduction and hip hiking [12], which was calculated by subtracting the maximum thigh angle in the coronal plane (i.e., the angle of the longitudinal axis of the thigh in the coronal plane in the global coordinate system) during 25–75% of the swing phase by the same time-averaged angle during the pre-swing phase (Figure 4e) [12,13,38].

2.10. *Statistical Analysis*

We initially performed a statistical analysis on a combined sample that integrated all three gait conditions to investigate the concurrent validity of the depth-sensor-based 3DMA in a sample that included individuals with and without compensatory movements for insufficient toe clearance (i.e., a population with high variability). Then, we examined the concurrent validity of the depth-sensor-based 3DMA in a sample of individuals with compensatory movements for insufficient toe clearance (i.e., a population with low variability) by performing statistical analyses for only the KAFO condition. Similarly, statistical analyses were conducted on the samples in each of the slow and very slow conditions (Supplementary Tables S1–S4).

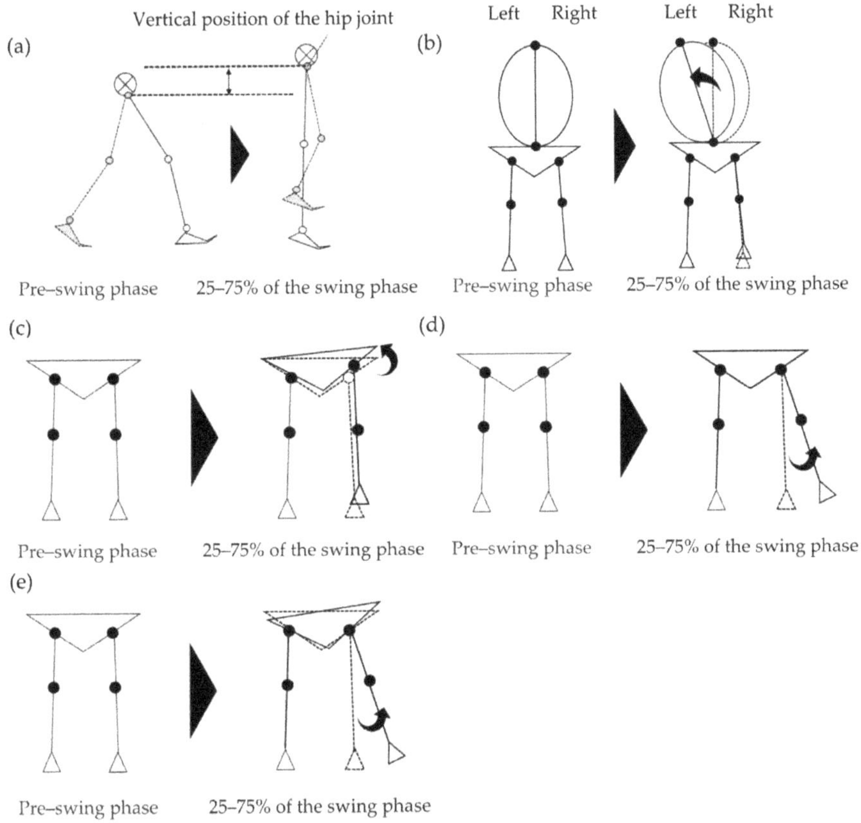

Figure 4. Definition of abnormal gait patterns: contralateral vaulting (**a**), trunk lateral flexion to the stance side (**b**), hip hiking (**c**), swing-side hip abduction (**d**), and circumduction (**e**).

The intraclass correlation coefficient (ICC) is the method for concurrent validation between the two gait analysis systems [19,39]. $ICC_{2,1}$ is an absolute agreement measure for evaluating the presence of an agreement between the two systems without taking into account the systematic errors, whereas $ICC_{3,1}$ is a relative agreement measure for evaluating the presence of an agreement between the two systems, taking into account the systematic errors. Therefore, we used $ICC_{2,1}$ to test whether the depth-sensor- and marker-based 3DMA results are in perfect agreement, and $ICC_{3,1}$ to test whether the depth-sensor-based 3DMA is an alternative to the marker-based 3DMA when systematic errors are taken into account. The ICC was classified as either poor (<0.50), moderate (0.50–0.74), good (0.75–0.90), or excellent (>0.90) [39]. The differences between the marker- and depth-sensor-based 3DMA were confirmed by performing paired *t*-tests. Additionally, the root mean square error (RMSE) between the gait parameters calculated with the marker- and depth-sensor-based 3DMA was identified [40]. IBM Statistical Package for the Social Sciences Statistics (version 26.0, IBM Corporation, Armonk, NY, USA) was used for all statistical analyses. In the present study, we chose "Intraclass correlation coefficient" under [Analyze] > [Scale] > [Reliability Analysis]. For $ICC_{2,1}$, the model was set to "Two-Way Random" and the type to "Absolute Agreement"; for $ICC_{3,1}$, these were "Two-Way Mixed" and "Consistency", respectively.

3. Results

3.1. Walking Speed in Each Condition

The mean and standard deviation of walking speeds were 0.64 ± 0.12, 0.38 ± 0.08, and 0.35 ± 0.07 m/s for the slow, very slow, and KAFO conditions, respectively, in the marker-based 3DMA.

3.2. Concurrent Validity of Abnormal Gait Patterns

In the combined sample, the $ICC_{3,1}$ of trunk lateral flexion and circumduction showed good validity, ranging from 0.76 to 0.85, whereas that of contralateral vaulting, hip hiking, and swing-side hip abduction showed moderate validity, ranging from 0.57 to 0.64 (Table 3). Conversely, the $ICC_{3,1}$ of hip hiking in the KAFO condition was 0.38, indicating poor validity. The $ICC_{2,1}$ results were generally slightly lower than the $ICC_{3,1}$ results. The measurement errors between instruments were 0.01 m, 1.30°, 1.99°, 2.37°, and 1.53° for contralateral vaulting, trunk lateral flexion, hip hiking, swing-side hip abduction, and circumduction, respectively.

Table 3. Concurrent validity of abnormal gait pattern.

Variables	Marker-Based 3DMA	Depth-Sensor-Based 3DMA	*p*-Value	$ICC_{2,1}$ (95% CI)	$ICC_{3,1}$ (95% CI)	RMSE
Combined sample						
Contralateral vaulting (m)	0.01 ± 0.01	0.01 ± 0.01	0.827	0.57 (0.37–0.72)	0.57 (0.37–0.71)	0.01
Trunk lateral flexion to stance side (deg)	3.39 ± 2.42	2.59 ± 2.06	0.624	0.80 (0.55–0.90)	0.85 (0.77–0.90)	1.30
Hip hiking (deg)	3.97 ± 4.87	2.91 ± 2.45	0.103	0.59 (0.39–0.72)	0.64 (0.47–0.77)	1.99
Swing-side hip abduction (deg)	2.47 ± 3.13	2.68 ± 1.92	0.498	0.59 (0.40–0.73)	0.59 (0.40–0.73)	2.37
Circumduction (deg)	4.01 ± 2.68	8.46 ± 2.57	<0.001 [a]	0.61 (0.26–0.78)	0.76 (0.63–0.84)	1.53
KAFO condition						
Contralateral vaulting (m)	0.01 ± 0.01	0.01 ± 0.01	0.644	0.51 (0.08–0.79)	0.56 (0.12–0.82)	0.01
Trunk lateral flexion to stance side (deg)	6.23 ± 2.31	4.82 ± 1.95	<0.001 [a]	0.66 (0.04–0.89)	0.79 (0.51–0.92)	1.43
Hip hiking (deg)	11.33 ± 1.90	5.80 ± 2.25	0.047 [a]	0.370.09–0.75)	0.380.11–0.72)	1.77
Swing-side hip abduction (deg)	1.09 ± 1.90	1.31 ± 2.17	0.083	0.510.09–0.84)	0.83 (0.60–0.94)	2.67
Circumduction (deg)	6.82 ± 2.77	10.43 ± 3.29	0.016 [a]	0.650.03–0.89)	0.72 (0.38–0.89)	0.96

Mean ± standard deviation; 3DMA: three-dimensional motion analysis; ICC: intraclass correlation coefficient; RMSE: root mean square error; KAFO: knee–ankle–foot orthosis; [a] significant difference between systems ($p < 0.05$).

3.3. Concurrent Validity of Spatiotemporal Variables

The $ICC_{2,1}$ and $ICC_{3,1}$ for all spatiotemporal variables, except %stance time, %swing time, and step length ratio, ranged from 0.77 to 0.99 in the combined sample, indicating good or excellent validity (Table 4). The other parameters (i.e., %stance time, %swing time, and step length ratio) demonstrated moderate validity ($ICC_{2,1}$ = 0.58–0.71, $ICC_{3,1}$ = 0.63–0.71; Table 4).

3.4. Concurrent Validity of Trunk and Lower Limb Kinematics

The average hip, knee, and ankle joint angles in the sagittal plane during the gait cycle evaluated by marker-based and depth-sensor-based 3DMA are shown in Figure 5. The $ICC_{3,1}$ for peak-to-peak angles of the trunk ranged from 0.76 to 0.85 in the combined sample, indicating good validity (Table 5). In the mean angle of the trunk, the $ICC_{3,1}$ for flexion–extension was good, and that for lateral flexion and rotation was moderate. The $ICC_{3,1}$ results revealed that the concurrent validity of the hip and knee joint-related parameters and leg extension angle was good or excellent ($ICC_{3,1}$ = 0.75–0.93), whereas the measure of absolute agreement was varied ($ICC_{2,1}$ = 0.23–0.93).

Table 4. Concurrent validity of spatiotemporal variables.

Variables		Marker-Based 3DMA	Depth-Sensor-Based 3DMA	p-Value	$ICC_{2,1}$ (95% CI)	$ICC_{3,1}$ (95% CI)	RMSE
Combined sample							
Step width (m)		0.18 ± 0.04	0.18 ± 0.04	0.213	0.89 (0.79–0.91)	0.87 (0.78–0.91)	0.02
Step length (m)	Right	0.34 ± 0.08	0.31 ± 0.08	<0.001 [a]	0.93 (0.51–0.98)	0.96 (0.95–0.98)	0.03
	Left	0.34 ± 0.08	0.32 ± 0.08	<0.001 [a]	0.93 (0.56–0.97)	0.96 (0.94–0.98)	0.03
Stride length (m)		0.75 ± 0.19	0.77 ± 0.19	<0.001 [a]	0.98 (0.96–0.99)	0.98 (0.98–0.99)	0.04
Cadence (steps/min)		78.17 ± 9.93	78.62 ± 10.20	0.146	0.97 (0.95–0.98)	0.97 (0.95–0.98)	2.40
Gait velocity (m/s)		0.47 ± 0.17	0.47 ± 0.17	0.013 [a]	0.99 (0.99–1.00)	0.99 (0.99–1.00)	0.01
Stance time (s)		1.03 ± 0.15	1.01 ± 0.15	0.222	0.93 (0.88–0.96)	0.94 (0.90–0.96)	0.05
Swing time (s)		0.52 ± 0.05	0.53 ± 0.05	0.477	0.80 (0.67–0.88)	0.82 (0.73–0.83)	0.03
Gait cycle time (s)		1.55 ± 0.19	1.54 ± 0.19	0.217	0.97 (0.96–0.98)	0.97 (0.95–0.98)	0.04
%Stance time (%)		66.5 ± 2.44	65.5 ± 2.33	<0.001 [a]	0.58 (0.34–0.74)	0.63 (0.45–0.76)	2.27
%Swing time (%)		33.5 ± 2.46	34.5 ± 2.37	<0.001 [a]	0.58 (0.34–0.74)	0.63 (0.43–0.76)	2.27
Step length ratio		0.99 ± 0.12	0.99 ± 0.15	0.705	0.71 (0.56–0.82)	0.71 (0.55–0.81)	0.10
Swing time ratio		1.06 ± 0.09	1.02 ± 0.10	0.021 [a]	0.71 (0.42–0.85)	0.77 (0.65–0.86)	0.08
KAFO condition							
Step width (m)		0.22 ± 0.03	0.22 ± 0.05	0.087	0.87(0.67–0.95)	0.88 (0.71–0.96)	0.02
Step length (m)	Right	0.28 ± 0.07	0.25 ± 0.06	<0.001 [a]	0.85 (0.17–0.96)	0.93 (0.81–0.97)	0.04
	Left	0.29 ± 0.06	0.28 ± 0.05	<0.001 [a]	0.93 (0.80–0.97)	0.94 (0.84–0.98)	0.03
Stride length (m)		0.62 ± 0.13	0.63 ± 0.11	0.121	0.97 (0.92–0.99)	0.97 (0.92–0.99)	0.03
Cadence (steps/min)		70.12 ± 1.54	71.03 ± 2.38	0.124	0.350.09–0.70)	0.740.11–0.72)	2.42
Gait velocity (m/s)		0.35 ± 0.07	0.35 ± 0.06	0.063	0.94 (0.90–0.99)	0.97 (0.92–0.99)	0.02
Stance time (s)		1.14 ± 0.05	1.13 ± 0.06	0.251	0.360.10–0.71)	0.370.12–0.71)	0.06
Swing time (s)		0.57 ± 0.05	0.57 ± 0.04	0.930	0.72 (0.38–0.89)	0.72 (0.37–0.86)	0.04
Gait cycle time (s)		1.71 ± 0.04	1.70 ± 0.05	0.180	0.350.11–0.70)	0.360.13–0.71)	0.04
%Stance time (%)		66.7 ± 2.60	66.3 ± 2.40	0.518	0.63 (0.22–0.84)	0.62 (0.21–0.84)	2.26
%Swing time (%)		33.3 ± 2.60	33.7 ± 2.40	0.518	0.63 (0.22–0.85)	0.62 (0.21–0.84)	2.26
Step length ratio		0.97 ± 0.15	0.91 ± 0.16	0.027 [a]	0.75 (0.38–0.91)	0.79 (0.52–0.92)	0.12
Swing time ratio		1.16 ± 0.11	1.12 ± 0.11	0.061	0.66 (0.28–0.86)	0.69 (0.35–0.88)	0.10

Mean ± standard deviation; 3DMA: three-dimensional motion analysis; ICC: intraclass correlation coefficients; RMSE: root mean square error; KAFO: knee–ankle–foot orthosis; [a] significant difference between systems ($p < 0.05$).

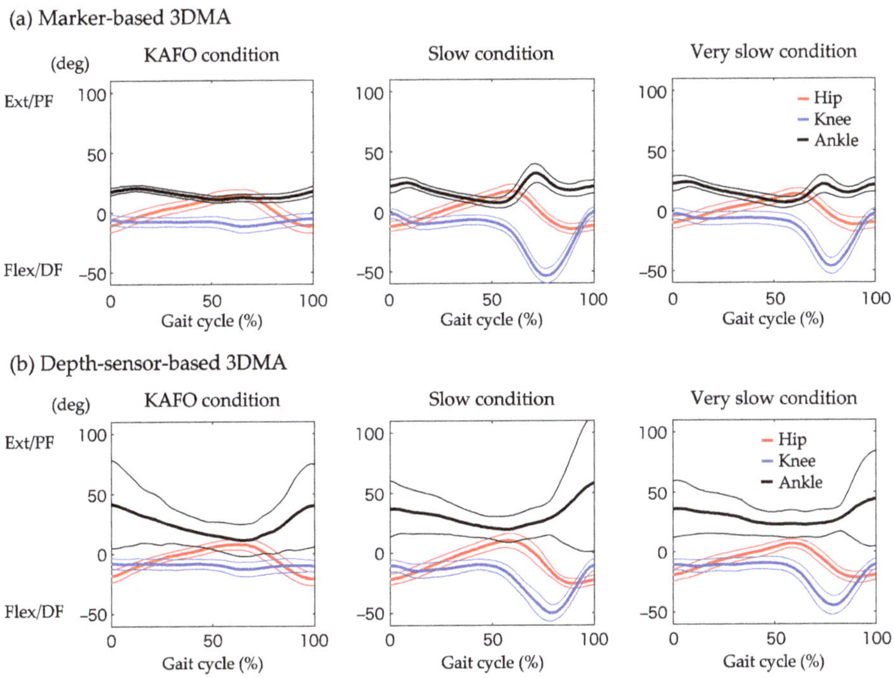

Figure 5. Average (thick line) and 1 SD (fine line) of hip and knee extension and ankle plantar flexion angles during gait cycle for each condition as evaluated with marker-based (**a**) and depth-sensor-based 3DMA (**b**) systems.

Table 5. Concurrent validity of the trunk and lower limb kinematics.

Variables	Marker-Based 3DMA	Depth-Sensor-Based 3DMA	*p*-Value	$ICC_{2,1}$ (95% CI)	$ICC_{3,1}$ (95% CI)	RMSE
Combined sample						
Trunk angle (deg)						
Peak-to-peak						
Flexion–extension	3.78 ± 1.77	4.73 ± 1.85	<0.001 [a]	0.74 (0.22–0.90)	0.85 (0.75–0.90)	1.49
Lateral flexion	5.89 ± 3.37	5.96 ± 2.13	0.099	0.76 (0.63–0.86)	0.76 (0.63–0.84)	1.89
Rotation	10.82 ± 3.45	13.62 ± 4.30	<0.001 [a]	0.79 (0.01–0.85)	0.79 (0.68–0.87)	3.77
Mean						
Flexion–extension	3.20 ± 3.20	2.70 ± 2.43	0.086	0.270.03–0.65)	0.85 (0.76–0.91)	6.10
Lateral flexion	0.71 ± 1.41	0.28 ± 1.06	<0.001 [a]	0.65 (0.44–0.77)	0.69 (0.53–0.80)	1.08
Rotation	−0.61 ± 3.87	−0.90 ± 3.17	0.386	0.73 (0.58–0.83)	0.72 (0.58–0.83)	2.63
Hip joint angle [+extension/−flexion] (deg)						
Maximum	15.89 ± 4.81	9.23 ± 4.63	<0.001 [a]	0.40 (0.07–0.75)	0.80 (0.69–0.87)	7.42
Minimum	−13.17 ± 4.94	−23.76 ± 4.46	<0.001 [a]	0.23 (0.03–0.60)	0.80 (0.70–0.88)	11.00
Peak-to-peak	29.05 ± 5.33	32.99 ± 6.23	0.092	0.72 (0.06–0.93)	0.93 (0.89–0.96)	4.45
Knee joint angle [+extension/−flexion] (deg)						
Maximum	−1.33 ± 4.18	−6.70 ± 2.75	<0.001 [a]	0.620.07–0.86)	0.75 (0.69–0.80)	6.24
Minimum	−40.04 ± 18.63	−39.86 ± 15.90	0.522	0.93 (0.83–0.95)	0.92 (0.88–0.95)	5.93
Peak-to-peak	38.71 ± 20.53	33.90 ± 15.10	<0.001 [a]	0.87 (0.70–0.93)	0.90 (0.84–0.94)	8.52
Ankle joint angle [+plantar flexion/−dorsal flexion] (deg)						
Maximum	28.05 ± 7.09	64.09 ± 33.10	<0.001 [a]	0.03 (−0.12–0.15)	0.05 (−0.30–0.20)	49.67
Minimum	6.61 ± 3.87	10.80 ± 8.56	<0.001 [a]	0.05(−0.12–0.10)	0.27 (−0.49–0.03)	10.81
Peak-to-peak	21.44 ± 8.06	53.29 ± 36.56	<0.001 [a]	0.43 (0.02–0.67)	0.57 (−0.37–0.71)	49.63
Maximum leg extension angle (deg)	19.82 ± 3.89	17.75 ± 3.74	<0.001 [a]	0.81 (0.03–0.93)	0.92 (0.87–0.95)	2.56
KAFO condition						
Trunk angle (deg)						
Peak-to-peak						
Flexion–extension	5.86 ± 1.94	7.00 ± 1.48	<0.001 [a]	0.71 (0.03–0.91)	0.85 (0.64–0.94)	1.47
Lateral flexion	10.31 ± 2.50	8.42 ± 1.65	<0.001 [a]	0.570.07–0.81)	0.69 (0.32–0.87)	1.89
Rotation	13.37 ± 3.84	15.92 ± 4.43	<0.001 [a]	0.69 (0.08–0.90)	0.81 (0.54–0.93)	3.77
Mean						
Flexion–extension	1.10 ± 3.40	4.67 ± 0.93	0.043 [a]	0.300.04–0.70)	0.83 (0.59–0.93)	6.02
Lateral flexion	1.56 ± 1.76	0.93 ± 1.30	0.048 [a]	0.67 (0.30–0.87)	0.71 (0.36–0.88)	1.34
Rotation	−1.98 ± 3.91	−2.50 ± 3.13	0.213	0.90 (0.74–0.96)	0.90 (0.74–0.96)	1.68
Hip joint angle [+extension/−flexion] (deg)						
Maximum	14.91 ± 4.55	8.58 ± 3.79	<0.001 [a]	0.360.08–0.75)	0.75 (0.43–0.90)	6.99
Minimum	−12.18 ± 5.13	−21.86 ± 4.55	<0.001 [a]	0.270.05–0.68)	0.76 (0.46–0.91)	10.24
Peak-to-peak	27.09 ± 3.75	30.44 ± 4.41	0.027 [a]	0.680.08–0.92)	0.90 (0.04–0.96)	3.84
Knee joint angle [+extension/−flexion] (deg)						
Maximum	−4.59 ± 4.14	−5.25 ± 3.21	<0.001 [a]	0.460.02–0.76)	0.450.02–0.75)	3.95
Minimum	−12.09 ± 4.94	−16.16 ± 4.29	<0.001 [a]	0.340.08–0.68)	0.450.02–0.76)	6.34
Peak-to-peak	7.49 ± 1.69	10.91 ± 2.83	<0.001 [a]	0.030.02–0.34)	0.070.41–0.52)	4.67
Ankle joint angle [+plantar flexion/−dorsal flexion] (deg)						
Maximum	21.30 ± 2.09	64.04 ± 25.50	0.033 [a]	0.020.12–0.28)	0.050.42–0.51)	49.46
Minimum	9.52 ± 1.84	4.42 ± 7.69	0.021 [a]	0.020.09–0.16)	0.060.52–0.41)	9.52
Peak-to-peak	11.78 ± 1.99	59.62 ± 29.52	<0.001 [a]	0.020.47–0.49)	0.020.45–0.46)	55.88
Maximum leg extension angle (deg)	17.78 ± 2.85	17.75 ± 3.74	<0.001 [a]	0.640.09–0.90)	0.85 (0.62–0.94)	2.70

Mean ± standard deviation; 3DMA: three-dimensional motion analysis; ICC: intraclass correlation coefficient; RMSE: root mean square error; KAFO: knee–ankle–foot orthosis; [a] significant difference between systems ($p < 0.05$).

3.5. Concurrent Validity of Whole-Body Kinematics

The $ICC_{2,1}$ and $ICC_{3,1}$ for the peak-to-peak value of CoM movement in the mediolateral direction ranged from 0.94 to 0.95 in the combined sample, indicating excellent validity (Table 6), whereas the concurrent validity of the peak-to-peak value of CoM movement in the vertical direction was poor. The $ICC_{3,1}$ of the MoS in the mediolateral and anteroposterior directions was 0.75–0.77, indicating good validity (Table 6).

Table 6. Concurrent validity of whole-body kinematics.

Variables	Marker-Based 3DMA	Depth-Sensor-Based 3DMA	*p*-Value	$ICC_{2,1}$ (95% CI)	$ICC_{3,1}$ (95% CI)	RMSE
Combined sample						
Peak-to-peak value of center of mass movement (m)						
Mediolateral	0.08 ± 0.03	0.08 ± 0.02	0.063	0.94 (0.91–0.97)	0.95 (0.91–0.97)	0.01
Vertical	0.05 ± 0.02	0.03 ± 0.01	<0.001 [a]	0.010.15–0.17)	0.110.26–0.26)	0.03
Mediolateral margin of stability (m)	0.09 ± 0.02	0.12 ± 0.03	<0.001 [a]	0.440.10–0.76)	0.77 (0.64–0.86)	0.04
Anteroposterior margin of stability (m)	−0.02 ± 0.03	0.08 ± 0.05	<0.001 [a]	0.210.04–0.56)	0.75 (0.61–0.84)	0.10
KAFO condition						
Peak-to-peak value of center of mass movement (m)						
Mediolateral	0.11 ± 0.02	0.11 ± 0.02	0.392	0.84 (0.55–0.94)	0.87 (0.70–0.94)	0.01
Vertical	0.05 ± 0.02	0.03 ± 0.01	0.013 [a]	0.100.38–0.30)	0.140.57–0.35)	0.03
Mediolateral margin of stability (m)	0.11 ± 0.03	0.14 ± 0.03	<0.001 [a]	0.460.11–0.80)	0.72 (0.38–0.89)	0.04
Anteroposterior margin of stability (m)	−0.02 ± 0.02	0.06 ± 0.03	<0.001 [a]	0.070.04–0.31)	0.410.07–0.74)	0.09

Mean ± standard deviation; 3DMA: three-dimensional motion analysis; ICC: intraclass correlation coefficient; RMSE: root mean square error; KAFO: knee–ankle–foot orthosis; [a] significant difference between systems ($p < 0.05$).

4. Discussion

The present study investigated the validity of the quantification of abnormal gait patterns (i.e., compensatory movements due to insufficient toe clearance) by depth-sensor-based 3DMA, as compared to marker-based 3DMA. The results of our study revealed good validity for trunk lateral flexion to the stance side and circumduction and moderate validity for contralateral vaulting, hip hiking, and swing-side hip abduction (Figure 6). The measurement error between instruments ranged from 1.30° to 2.37° and was 0.01 m for contralateral vaulting. Most previous studies have validated depth-sensor-based 3DMA for lower limb joint angles in the sagittal plane, which revealed measurement errors of 4.1–11.3°, 5.3–8.6°, 9.14–38.6°, and 1.11–5.12° for maximum hip flexion, hip extension, knee joint flexion, and knee joint extension angles, respectively, during gait [40,41]. The measurement error of abnormal gait patterns in the coronal plane reported in this study is very small compared to the results of previous studies. This is not a surprising result, because the depth-sensor-based 3DMA estimates the 3D coordinate values of the body landmarks using the contour and depth information of a subject captured from the front. In a previous study investigating the validity of depth-sensor-based 3DMA for estimating hip motion in the coronal plane, a measurement error between devices of 12.5° in hip abduction angle throughout a gait cycle was reported [42]. The present study used Azure Kinect, the successor to Kinect v2 used in the previous study. The smaller measurement error in the abnormal gait pattern measures in the coronal plane in the present study, as compared to that of the previous study, may indicate the higher performance of Azure Kinect's body landmark estimation. To the best of our knowledge, this is the first study to reveal that depth-sensor-based 3DMA is useful for quantifying compensatory movements due to insufficient toe clearance during the swing phase of gait. However, the circumduction evaluated by depth-sensor-based 3DMA was significantly higher than that by marker-based-3DMA. It must be considered that depth-sensor-based 3DMA estimates body landmarks rather than measuring them; thus, it does not perfectly match the systems that actually measure marker locations.

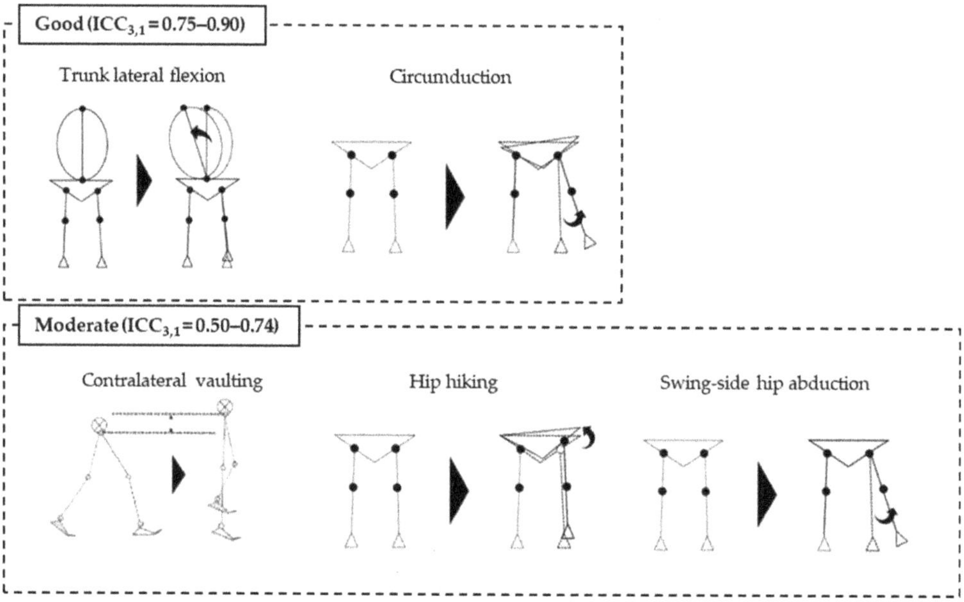

Figure 6. Summary for the concurrent validity of abnormal gait patterns.

The $ICC_{3,1}$ demonstrated good or excellent validity for most of the spatiotemporal and kinematic variables, supporting the results of previous studies [18,19,32,43]. The measurement errors of the angles of the trunk and lower extremity joints, excluding the ankle joint, ranged from 1.08° to 11.0°, which was equal to or less than the results of the aforementioned previous studies [40,41]. Additionally, comparisons between systems showed significantly shorter step lengths, significantly longer %swing phase time, and significantly greater hip and knee flexion for the depth-sensor-based 3DMA than for the marker-based 3DMA. When comparing the results between the marker- and depth-sensor-based 3DMA, we should consider the measurement error. Only the peak-to-peak value of the CoM movement in the vertical direction exhibited poor validity, which is inconsistent with the result of a previous study [24]. Ferraris et al. [24] revealed a highly valid CoM vertical sway, but they defined the CoM position as the midpoint between the left and right hip joints. The CoM is generally calculated from the position and mass of each body segment, and the CoM position described in previous studies was based on a simplified definition. Methodological differences may have caused discrepancies between the results of the previous study and this study. Our study results reveal a greater measurement error of kinematic variables in the sagittal plane than in the coronal plane. Therefore, the peak-to-peak value of the vertical CoM movement may have had less validity because the flexion–extension movement of each joint in the sagittal plane primarily influenced it.

The statistical analysis for the combined sample revealed moderate validity of the assessment for hip hiking using depth-sensor-based 3DMA, but it is important to consider that the greater the variability of the data in the collected sample, the higher the reliability coefficient [44]. Taking this issue into account, the present study conducted a statistical analysis for only the KAFO condition, which demonstrated compensatory movement due to insufficient toe clearance during the swing phase. Hence, the validity of the hip hiking measurement was poor. The hip hiking evaluated by the marker-based 3DMA revealed 11.30° ± 1.90° (mean ± standard deviation) and 3.97° ± 4.87° in the KAFO condition and combined sample, respectively. Therefore, the differences in the standard deviations between conditions may affect the validity results. However, a previous study reported hip hiking of 8.7° ± 4.3° in 23 post-stroke patients [11], which is comparable to the standard deviation of hip hiking in the combined sample of our study. The depth-sensor-based 3DMA may be useful for identifying the compensatory movements during the swing phase, including hip hiking, in post-stroke patients, considering the results of the present study and previous studies, despite the possibility of missing a small difference of approximately 2° of measurement error.

The high validity of depth-sensor-based 3DMA, i.e., markerless 3DMA, for quantifying compensatory movements may accelerate the accumulation of data on gait characteristics in clinical practice. Kinematic and neurological factors reportedly cause compensatory movements, including circumduction and hip hiking, in post-stroke patients [45,46]. Inadequate knee joint flexion during the pre-swing phase of the paretic side is the trigger in the case of kinetic factors, whereas the co-activation of the gluteus medius, which occurs simultaneously with an abnormal stretch reflex in the rectus femoris muscle, is the trigger in the case of neurological factors. As described previously, only a few longitudinal data provide a basis for the effective interventions for compensatory movements during the swing phase in post-stroke patients, despite the partially clarified triggers of compensatory movements among these patients. Accumulating changes in compensatory movements due to insufficient toe clearance before and after a rehabilitation intervention using depth-sensor-based 3DMA may be useful in providing effective gait rehabilitation for post-stroke patients.

The present study has several limitations. First, we defined cases in which body landmark estimation could not be conducted using depth-sensor-based 3DMA as failed trials and excluded them from the data analysis. Prior to the actual experiment, we noticed that the errors in skeletal estimation occur when the reflective markers are large and numerous. Thus, we adjusted the number and size of the markers as described in the Materials and Methods section. Despite this countermeasure, several failed trials still occurred. Unfortunately, our results revealed no reason for the failure. Although we expect the failure rate to decrease in clinical settings where reflective markers are not attached, further investigation is still required to determine the interfering factors, such as the participants' body types, wear, ambient lighting, and other environmental conditions. Second, the effect of clothing on determining the validity of depth-sensor-based 3DMA was not considered. All of our study participants wore spandex clothing, which may have increased the study's validity. Restricting clothing is necessary when conducting gait assessment with a depth-sensor-based 3DMA in clinical practice. Third, the effect of the presence or absence of a caregiver on 3D coordinate data by depth-sensor-based 3DMA has not been investigated. This study captured and analyzed only the participants undergoing gait assessment. Placing a caregiver near the patient is often necessary to prevent falls when the depth-sensor-based 3DMA is used in clinical practice. Therefore, investigating the association of the presence of a caregiver with difficulties in measuring 3D coordinate data during gait is essential. Fourth, the present study did not include post-stroke patients. However, depth-sensor-based 3DMA was effective in quantifying compensatory movements during the swing phase, as we examined the validity based on the abnormal gait patterns during the swing phase induced by the KAFO. Finally, this study did not consider simultaneous capture from the sagittal plane. Combining imaging from the sagittal plane may solve the problem of ankle joint motion evaluation, which was very inaccurate in our study.

5. Conclusions

We revealed the high concurrent validity of quantifying compensatory movements due to insufficient toe clearance using a depth-sensor-based 3DMA. The gait assessment using a depth-sensor-based 3DMA can potentially provide a rehabilitation program based on the objective evaluation of gait in clinical practice.

Supplementary Materials: The following supporting information can be downloaded at https://www.mdpi.com/article/10.3390/biomechanics4030028/s1: Table S1: The concurrent validity of the abnormal gait patterns in the slow and very slow conditions, Table S2: The concurrent validity of the spatiotemporal variables in the slow and very slow conditions, Table S3: The concurrent validity of the trunk and lower limb kinematics in the slow and very slow conditions, Table S4: The concurrent validity of the whole-body kinematics in the slow and very slow conditions.

Author Contributions: Conceptualization, K.K., S.M., D.N., M.K., H.H., T.M. (Toshihiro Miwa), T.M. (Toru Maeda), T.N., S.N. and K.H.; Data curation, K.K., D.N., T.M. (Toshihiro Miwa), and K.H.; Formal analysis, K.K. and K.H.; Funding acquisition, K.H.; Investigation, K.K. and K.H.; Methodology, K.K., S.M., D.N., M.K., H.H., T.M. (Toshihiro Miwa), T.M. (Toru Maeda), T.N., S.N. and K.H.; Project administration, K.K. and K.H.; Resources, K.H.; Software, K.K. and K.H.; Supervision, S.N. and K.H.; Validation, K.K. and K.H.; Visualization, K.K. and K.H.; Writing—original draft, K.K. and K.H.; Writing—review and editing, K.K., S.M., D.N., M.K., H.H., T.M. (Toshihiro Miwa), T.M. (Toru Maeda), T.N., S.N. and K.H. All authors have read and agreed to the published version of the manuscript.

Funding: This work was supported by JSPS KAKENHI [Grant Number 22K11443].

Institutional Review Board Statement: The study was conducted in accordance with the guidelines stipulated in the Declaration of Helsinki and approved by the Ethics Committee of Kumamoto Health Science University (protocol code: 22043; date of approval: 7 February 2023).

Informed Consent Statement: Informed consent was obtained from all participants involved in the study. Written informed consent was obtained from the participants to publish this paper.

Data Availability Statement: Data are available on request due to privacy and ethical restrictions.

Acknowledgments: We would like to thank TOKUDA Prosthetics and Orthotics Mfg Ltd. for lending us the Modular Leg Brace NEO.

Conflicts of Interest: The authors have received KAFO (Modular Leg Brace NEO) on loan from TOKUDA Prosthetics and Orthotics Mfg Ltd.

References

1. Van de Port, I.G.; Kwakkel, G.; Lindeman, E. Community ambulation in patients with chronic stroke: How is it related to gai speed? *J. Rehabil. Med.* **2008**, *40*, 23–27. [CrossRef] [PubMed]
2. Thilarajah, S.; Mentiplay, B.F.; Bower, K.J.; Tan, D.; Pua, Y.H.; Williams, G.; Koh, G.; Clark, R.A. Factors associated with post-stroke physical activity: A systematic review and meta-analysis. *Arch. Phys. Med. Rehabil.* **2018**, *99*, 1876–1889. [CrossRef] [PubMed]
3. Weerdesteyn, V.; de Niet, M.; van Duijnhoven, H.J.; Geurts, A.C. Falls in individuals with stroke. *J. Rehabil. Res. Dev.* **2008**, *45* 1195–1213. [CrossRef] [PubMed]
4. Michael, K.M.; Allen, J.K.; Macko, R.F. Reduced ambulatory activity after stroke: The role of balance, gait, and cardiovascula fitness. *Arch. Phys. Med. Rehabil.* **2005**, *86*, 1552–1556. [CrossRef] [PubMed]
5. Perry, J.; Garrett, M.; Gronley, J.K.; Mulroy, S.J. Classification of walking handicap in the stroke population. *Stroke* **1995**, *26* 982–989. [CrossRef] [PubMed]
6. Fulk, G.D.; He, Y.; Boyne, P.; Dunning, K. Predicting home and community walking activity poststroke. *Stroke* **2017**, *48*, 406–411 [CrossRef] [PubMed]
7. Hsiao, H.; Awad, L.N.; Palmer, J.A.; Higginson, J.S.; Binder-Macleod, S.A. Contribution of paretic and nonparetic limb peak propulsive forces to changes in walking speed in individuals poststroke. *Neurorehabil. Neural Repair* **2016**, *30*, 743–752. [CrossRef] [PubMed]
8. Patterson, K.K.; Mansfield, A.; Biasin, L.; Brunton, K.; Inness, E.L.; McIlroy, W.E. Longitudinal changes in poststroke spatiotemporal gait asymmetry over inpatient rehabilitation. *Neurorehabil. Neural Repair* **2015**, *29*, 153–162. [CrossRef] [PubMed]
9. Awad, L.N.; Palmer, J.A.; Pohlig, R.T.; Binder-Macleod, S.A.; Reisman, D.S. Walking speed and step length asymmetry modify the energy cost of walking after stroke. *Neurorehabil. Neural Repair* **2015**, *29*, 416–423. [CrossRef]
10. Jørgensen, L.; Crabtree, N.J.; Reeve, J.; Jacobsen, B.K. Ambulatory level and asymmetrical weight bearing after stroke affects bone loss in the upper and lower part of the femoral neck differently: Bone adaptation after decreased mechanical loading. *Bone* **2000** *27*, 701–707. [CrossRef]
11. Kerrigan, D.C.; Frates, E.P.; Rogan, S.; Riley, P.O. Hip hiking and circumduction: Quantitative definitions. *Am. J. Phys. Med Rehabil.* **2000**, *79*, 247–252. [CrossRef]
12. Stanhope, V.A.; Knarr, B.A.; Reisman, D.S.; Higginson, J.S. Frontal plane compensatory strategies associated with self-selected walking speed in individuals post-stroke. *Clin. Biomech.* **2014**, *29*, 518–522. [CrossRef]
13. Tyrell, C.M.; Roos, M.A.; Rudolph, K.S.; Reisman, D.S. Influence of systematic increases in treadmill walking speed on gait kinematics after stroke. *Phys. Ther.* **2011**, *91*, 392–403. [CrossRef] [PubMed]
14. Mukaino, M.; Ohtsuka, K.; Tanikawa, H.; Matsuda, F.; Yamada, J.; Itoh, N.; Saitoh, E. Clinical-oriented three-dimensional gai analysis method for evaluating gait disorder. *J. Vis. Exp.* **2018**, *133*, e57063. [CrossRef]

15. Clark, R.A.; Vernon, S.; Mentiplay, B.F.; Miller, K.J.; McGinley, J.L.; Pua, Y.H.; Paterson, K.; Bower, K.J. Instrumenting gait assessment using the Kinect in people living with stroke: Reliability and association with balance tests. *J. Neuroeng. Rehabil.* **2015**, *12*, 15. [CrossRef] [PubMed]
16. Latorre, J.; Colomer, C.; Alcañiz, M.; Llorens, R. Gait analysis with the Kinect v2: Normative study with healthy individuals and comprehensive study of its sensitivity, validity, and reliability in individuals with stroke. *J. Neuroeng. Rehabil.* **2019**, *16*, 97. [CrossRef]
17. Lachat, E.; Macher, H.; Landes, T.; Grussenmeyer, P. Assessment and calibration of a RGB-D camera (Kinect v2 Sensor) towards a potential use for close-range 3D modeling. *Remote Sens.* **2015**, *7*, 13070–13097. [CrossRef]
18. Usami, T.; Nishida, K.; Iguchi, H.; Okumura, T.; Sakai, H.; Ida, R.; Horiba, M.; Kashima, S.; Sahashi, K.; Asai, H.; et al. Evaluation of lower extremity gait analysis using Kinect V2® tracking system. *SICOT J.* **2022**, *8*, 27. [CrossRef]
19. Eltoukhy, M.; Oh, J.; Kuenze, C.; Signorile, J. Improved kinect-based spatiotemporal and kinematic treadmill gait assessment. *Gait Posture* **2017**, *51*, 77–83. [CrossRef]
20. Matsuda, F.; Mukaino, M.; Ohtsuka, K.; Tanikawa, H.; Tsuchiyama, K.; Teranishi, T.; Kanada, Y.; Kagaya, H.; Saitoh, E. Biomechanical factors behind toe clearance during the swing phase in hemiparetic patients. *Top. Stroke Rehabil.* **2017**, *24*, 177–182. [CrossRef]
21. Zissimopoulos, A.; Fatone, S.; Gard, S.A. Biomechanical and energetic effects of a stance-control orthotic knee joint. *J. Rehabil. Res. Dev.* **2007**, *44*, 503–513. [CrossRef] [PubMed]
22. Akbas, T.; Prajapati, S.; Ziemnicki, D.; Tamma, P.; Gross, S.; Sulzer, J. Hip circumduction is not a compensation for reduced knee flexion angle during gait. *J. Biomech.* **2019**, *87*, 150–156. [CrossRef] [PubMed]
23. Wang, Y.; Mukaino, M.; Ohtsuka, K.; Otaka, Y.; Tanikawa, H.; Matsuda, F.; Tsuchiyama, K.; Yamada, J.; Saitoh, E. Gait characteristics of post-stroke hemiparetic patients with different walking speeds. *Int. J. Rehabil. Res.* **2020**, *43*, 69–75. [CrossRef]
24. Ferraris, C.; Cimolin, V.; Vismara, L.; Votta, V.; Amprimo, G.; Cremascoli, R.; Galli, M.; Nerino, R.; Mauro, A.; Priano, L. Monitoring of gait parameters in post-stroke individuals: A feasibility study using RGB-D sensors. *Sensors* **2021**, *21*, 5945. [CrossRef] [PubMed]
25. Learn Microsoft. Available online: https://learn.microsoft.com/ja-jp/azure/kinect-dk/body-joints (accessed on 14 May 2024).
26. Kurabayashi, J.; Mochimaru, M.; Kouchi, M. Validation of the estimation methods for the hip joint center. *J. Soc. Biomech.* **2003**, *27*, 28–36. [CrossRef]
27. Tokuda, K.; Anan, M.; Takahashi, M.; Sawada, T.; Tanimoto, K.; Kito, N.; Shinkoda, K. Biomechanical mechanism of lateral trunk lean gait for knee osteoarthritis patients. *J. Biomech.* **2018**, *66*, 10–17. [CrossRef]
28. Kagami, S.; Mochimaru, M.; Ehara, Y.; Miyata, N.; Nishiwaki, K.; Kanade, T.; Inoue, H. Measurement and comparison of humanoid H7 walking with human being. *Robot. Auton. Syst.* **2004**, *48*, 177–187. [CrossRef]
29. Mizuta, N.; Hasui, N.; Kai, T.; Inui, Y.; Sato, M.; Ohnishi, S.; Taguchi, J.; Nakatani, T. Characteristics of limb kinematics in the gait disorders of post-stroke patients. *Sci. Rep.* **2024**, *14*, 3082. [CrossRef] [PubMed]
30. Winter, D.A.; Quanbury, A.O.; Hobson, D.A.; Sidwall, H.G.; Reimer, G.; Trenholm, B.G.; Steinke, T.; Shlosser, H. Kinematics of normal locomotion—A statistical study based on T.V. data. *J. Biomech.* **1974**, *7*, 479–486. [CrossRef]
31. Clark, R.A.; Bower, K.J.; Mentiplay, B.F.; Paterson, K.; Pua, Y.H. Concurrent validity of the Microsoft Kinect for assessment of spatiotemporal gait variables. *J. Biomech.* **2013**, *46*, 2722–2725. [CrossRef]
32. Mentiplay, B.F.; Perraton, L.G.; Bower, K.J.; Pua, Y.H.; McGaw, R.; Heywood, S.; Clark, R.A. Gait assessment using the Microsoft Xbox One Kinect: Concurrent validity and inter-day reliability of spatiotemporal and kinematic variables. *J. Biomech.* **2015**, *48*, 2166–2170. [CrossRef] [PubMed]
33. French, M.A.; Koller, C.; Arch, E.S. Comparison of three kinematic gait event detection methods during overground and treadmill walking for individuals post stroke. *J. Biomech.* **2020**, *99*, 109481. [CrossRef] [PubMed]
34. Zeni, J.A., Jr.; Richards, J.G.; Higginson, J.S. Two simple methods for determining gait events during treadmill and overground walking using kinematic data. *Gait Posture* **2008**, *27*, 710–714. [CrossRef] [PubMed]
35. Matsuzawa, Y.; Miyazaki, T.; Takeshita, Y.; Higashi, N.; Hayashi, H.; Araki, S.; Nakatsuji, S.; Fukunaga, S.; Kawada, M.; Kiyama, R. Effect of leg extension angle on knee flexion angle during swing phase in post-stroke gait. *Medicina* **2021**, *57*, 1222. [CrossRef] [PubMed]
36. Hof, A.L.; Gazendam, M.G.; Sinke, W.E. The condition for dynamic stability. *J. Biomech.* **2005**, *38*, 1–8. [CrossRef] [PubMed]
37. Tanikawa, H.; Ohtsuka, K.; Mukaino, M.; Inagaki, K.; Matsuda, F.; Teranishi, T.; Kanada, Y.; Kagaya, H.; Saitoh, E. Quantitative assessment of retropulsion of the hip, excessive hip external rotation, and excessive lateral shift of the trunk over the unaffected side in hemiplegia using three-dimensional treadmill gait analysis. *Top. Stroke Rehabil.* **2016**, *23*, 311–317. [CrossRef]
38. Itoh, N.; Kagaya, H.; Saitoh, E.; Ohtsuka, K.; Yamada, J.; Tanikawa, H.; Tanabe, S.; Itoh, N.; Aoki, T.; Kanada, Y. Quantitative assessment of circumduction, hip hiking, and forefoot contact gait using lissajous figures. *Jpn. J. Compr. Rehabil. Sci.* **2012**, *3*, 78–84. [CrossRef]
39. Koo, T.K.; Li, M.Y. A Guideline of Selecting and Reporting Intraclass Correlation Coefficients for Reliability Research. *J. Chiropr. Med.* **2016**, *15*, 155–163. [CrossRef] [PubMed]
40. Pfister, A.; West, A.M.; Bronner, S.; Noah, J.A. Comparative abilities of Microsoft Kinect and Vicon 3D motion capture for gait analysis. *J. Med. Eng. Technol.* **2014**, *38*, 274–280. [CrossRef]

41. Xu, X.; McGorry, R.W.; Chou, L.S.; Lin, J.H.; Chang, C.C. Accuracy of the Microsoft Kinect for measuring gait parameters during treadmill walking. *Gait Posture* **2015**, *42*, 145–151. [CrossRef]
42. Ma, Y.; Mithraratne, K.; Wilson, N.C.; Wang, X.; Ma, Y.; Zhang, Y. The validity and reliability of a kinect v2-based gait analysis system for children with cerebral palsy. *Sensors* **2019**, *19*, 1660. [CrossRef]
43. Tamura, H.; Tanaka, R.; Kawanishi, H. Reliability of a markerless motion capture system to measure the trunk, hip and knee angle during walking on a flatland and a treadmill. *J. Biomech.* **2020**, *109*, 109929. [CrossRef]
44. Stratford, P.W.; Goldsmith, C.H. Use of the standard error as a reliability index of interest: An applied example using elbow flexor strength data. *Phys. Ther.* **1997**, *77*, 745–750. [CrossRef] [PubMed]
45. Akbas, T.; Neptune, R.R.; Sulzer, J. Neuromusculoskeletal simulation reveals abnormal rectus femoris-gluteus medius coupling in post-stroke gait. *Front. Neurol.* **2019**, *10*, 301. [CrossRef] [PubMed]
46. Hall, A.L.; Peterson, C.L.; Kautz, S.A.; Neptune, R.R. Relationships between muscle contributions to walking subtasks and functional walking status in persons with post-stroke hemiparesis. *Clin. Biomech.* **2011**, *26*, 509–515. [CrossRef] [PubMed]

biomechanics

MDPI

Article

Effects of Gait Speed and Sole Adjustment on Shoe–Floor Angles: Measurement Using Shoe-Type Sensor

Yu Hashiguchi [1,*], Tsuguru Numabe [2] and Ryosuke Goto [3]

1 Department of Physical Therapy, Faculty of Rehabilitation, Gunma Paz University, Takasaki 370-0006, Japan
2 Department of Rehabilitation, Yuai Orthopedics Clinic, Tomioka 370-2315, Japan
3 Department of Speech-Language-Hearing Therapy, Faculty of Rehabilitation, Gunma Paz University, Takasaki 370-0006, Japan
* Correspondence: hashiguchi@paz.ac.jp; Tel.: +81-90-7883-5292

Abstract: Background: Assessment of walking with shoes is important for understanding different types of walking in various environments. Methods: In this study, a shoe-type sensor was used to demonstrate the shoe–floor angle in fifteen participants who walked on a treadmill under varying gait speed and sole adjustments, lifting one side of the sole. The shoe–floor angle in the sagittal; the angle of toe-up (θTup) and toe-down (θTdown) and frontal planes; and the angle of pronation (θPro) and supination (θSup) were calculated, and angles at the initial contact and maximum angles were extracted. Results: The results showed that most angles significantly increased with an increase in the gait speed (θTup and θTdown; $p < 0.01$ both, θPro and θSup; $p < 0.02$ and 0.04). Conversely, only the supination angle at the initial contact changed significantly, owing to the tilt of the sole ($p < 0.01$). Conclusion: Shoe movements were more strongly affected by gait speed than by sole adjustment.

Keywords: gait analysis; shoe-type motion sensor; gait speed; sole adjustment

1. Introduction

Individuals wear shoes for a significant amount of time. Hence, the evaluation of walking in shoes is essential for understanding the different types of walking in various environments. Previous studies reported differences between barefoot and shoe-wearing gaits [1]. An increase in stride length [2] and a decrease in cadence [3] have been reported when wearing shoes. Furthermore, an increase in the anteroposterior center of pressure displacement has been reported [4]. The range of motion of the ankle joint is significantly increased by wearing shoes [4], and differences in the plantar flexion moment [5] and the propulsive ground reaction force [6] have also been observed. Shoes may increase the forefoot stiffness by decreasing the spread of the forefoot and providing support against the increased plantar flexion moment during the late stance phase. However, the effects of the gait conditions on shoe movements remain unclear. Therefore, clarifying the kinematic changes in gait conditions is needed for designing shoes in various environments.

Although various gait analysis methods have been reported, the simplest and easiest method is the use of an accelerometer. In particular, lumbar-mounted accelerometers are widely used for the quantification of gait [7]. Menz et al. [8] calculated the root mean square range of motion values of acceleration data measured using a lumbar accelerometer to evaluate the sway of the center of gravity. To show the periodicity of gait, Moe-Nilssen et al. [9] calculated the similarity of the acceleration waveforms between each gait cycle. Other methods that indicate the harmony and stationarity of gait, such as the harmonic ratio [10] and entropy [11], have also been reported. Subsequently, lower leg-mounted accelerometers were used to demonstrate gait events, such as heel contact, and the previous study showed the gait instability using the variability of stride time [12].

A method for measuring foot movements using a foot-mounted accelerometer has also been reported [13]. Previous studies have used the foot–floor angle (FFA), which is the angle between the foot and floor. FFAs in the sagittal plane are clinically important for gait speed [14], slip accidents [15], and falls [16]. The pronation angle in the frontal plane has been shown to be associated with low back pain [17] and injuries during running [18]. Although the characteristics of FFAs have been reported, the angle between the shoe and floor (shoe–floor angle; SFA) has not been clear. Furthermore, it requires the sensors to be placed inside the shoes to clarify the SFA directly.

In this study, the ORPHE, a shoe-type sensor with a built-in accelerometer, was used. The validity of the method using this sensor was clarified in a previous study [19]. ORPHE can directly measure and demonstrate the details of SFAs in real time. In this study, we evaluated the changes in the parameters measured using this sensor under various gait speed and sole adjustment conditions. The purpose of this study is to clarify the change in the SFA due to gait speed and sole adjustment. We hypothesized that changes in gait speed and shoe adjustment cause the change in the SFA in the sagittal plane and frontal plane, respectively. The results of this study are expected to provide basic data for gait analyses in diverse environments.

2. Materials and Methods

2.1. Participants

A total of 15 healthy young adults (11 males and 4 females) participated in the study (mean ± standard deviation: age 27.3 ± 9.1 years; height 172.0 ± 8.0 cm; weight 67.9 ± 12.1 kg). The sample size was calculated by G*Power 3.1 (effect size was set at 0.5, significance level at 5%, and power at 0.8) with reference to a previous study [20]. The exclusion criteria were that they had been diagnosed with a musculoskeletal or neurological disease within 2 months of the measurement and current treatment. Furthermore, the inclusion criterion was the absence of pain or instability during walking. The participants were faculty and students affiliated with medical universities who had a general standard of physical activity without excessive or special training.

2.2. Procedure

Gait measurements were performed while the participants wore shoes (SHIBUYA 2.0, ORPHE Inc., Tokyo, Japan) on a treadmill (AUTO RUNNER AR-200; MINATO MERDICAL SCIENCE Co., Osaka, Japan). The shoe-type sensor ORPHE TRACK (ORPHE Inc., Tokyo, Japan; size: 45 mm × 29 mm × 14 mm; weight: 20 g, sampling frequency: 200 Hz; Figure 1) measured the three-axial accelerations and angular velocities using a built-in sensor. Shoe sizes were selected from a range of 23 to 28 cm in 1 cm increments. Five walking conditions were performed: two conditions of gait speed change, two conditions of change in sole adjustments, and the normal condition. The treadmill was set at an inclined angle of 0°, and the walking speed for each condition was set before the measurement. The starting speed was set to 0 km/h and was gradually increased until the set speed was reached after a few seconds; after the set speed was reached, a constant speed was maintained. During the measurements, the participants walked for 1 min under each condition, and the data for the following 15 s were measured.

For the normal condition (Normal), the walking speed was set to 4 km/h; for the other two conditions, the walking speed was set to 3 km/h (Slow) and 5 km/h (Fast). The average speed during overground walking was 4.5–4.8 km/h [21]. However, a comfortable gait speed on a treadmill has been reported to be slower than that of overground walking [22]. To examine changes in speed, the other two conditions were set to ±1 km/h from the walking speed set for the Normal condition. For the sole adjustment condition, 2 cm wide and 6 mm thick ethylene-vinyl acetate hard sponges were cut from the calcaneus to the thenar and attached under the shoe insert with double-sided tape. In each condition, the sponges were inserted in the medial (Medial) and lateral (Lateral) positions to lift one side of the sole. Each condition was performed in a random order to minimize learning effects.

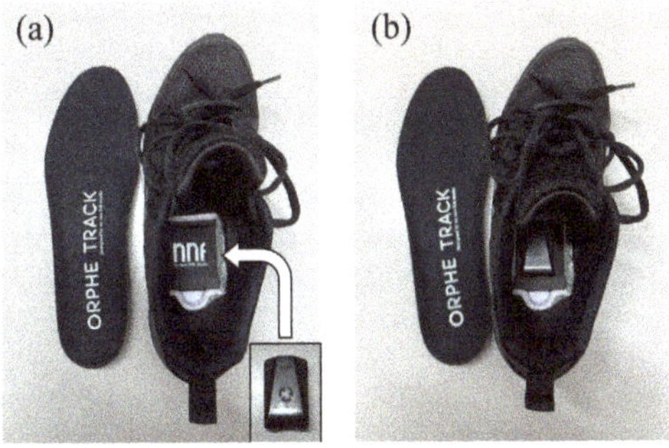

Figure 1. (**a**) A sensor embedded in the shoe midsole; (**b**) shoes with the sensor.

2.3. Data Analysis

The data obtained were analyzed using ORPHE ANALYTICS (ORPHE Inc.) [19]. Euler angles were calculated using a Madgwick filter [23] from the obtained acceleration and angular velocity. The SFAs were demonstrated on the sagittal and frontal planes. Data from five gait cycles were extracted and normalized to a 100% gait cycle, and the average waveform was calculated. As shown in Figure 2, the angle of toe-up (θTup) and toe-down (θTdown) in the sagittal plane, and the angle of supination (θSup) and pronation (θPro) in the frontal plane were demonstrated.

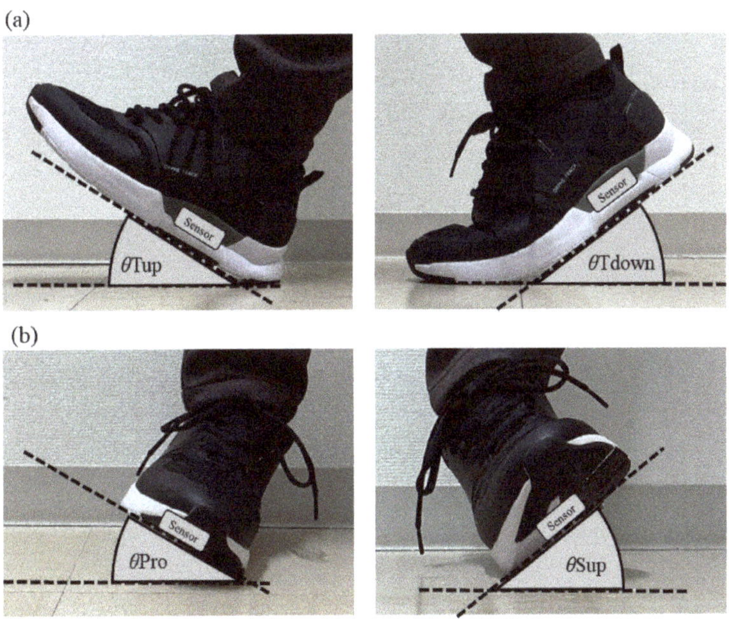

Figure 2. Parameters of shoe–floor angle. These parameters demonstrate the angle between the ground and shoe in the sagittal (**a**) or frontal plane (**b**) that were calculated through the sensor's anteroposterior or mediolateral axis and its projection vector to the horizontal plane. The angle of toe-up (θTup) and toe-down (θTdown) in the sagittal plane, and the angle of supination (θSup) and pronation (θPro) in the frontal plane were demonstrated.

For outcomes, θTup and θSup at the initial contact and the maximum θTdown and θPro were extracted as the magnitude of the SFA relating to important events during gait. The waveform obtained was expressed as 100% for each gait cycle and standardized across the participants. This approach enabled calculating the timing within the gait cycles, and the timing of maximum θTdown and θPro were evaluated as the timing of the SFA. The average and median values of parameters were calculated for each condition as a representative value.

2.4. Statistical Analysis

Statistical analyses were performed using the SPSS software (version 28; IBM, Armonk, NY, USA). The measured data were tested for normality using the Shapiro–Wilk test. The test showed that some of the data were not normally distributed. Table 1 shows the average and median value of parameters and the results of the Shapiro–Wilk test in each condition. To compare the parameters using a consistent method, we used the Friedman test for all parameters to examine the differences due to changes in gait speed and sole adjustment. In addition, a post-hoc analysis with the Bonferroni correction was performed to compare the conditions. The statistical significance was set at 5%.

Table 1. The average and median values of parameters among each condition and results of the Shapiro–Wilk test.

Parameters	Condition	Average (SD)	Median (IQR)	*p*-Value of Shapiro–Wilk Test
θTup at Initial contact				
	Normal	21.4 (6.3)	23.3 (18.6–23.9)	0.49
	Slow	17.3 (6.0)	18.8 (14.3–21.3)	0.54
	Fast	25.3 (6.3)	25.4 (20.8–28.9)	0.97
	Medial	21.4 (6.7)	24.0 (16.9–26.2)	0.10
	Lateral	19.1 (5.9)	21.8 (15.5–22.9)	0.04 *
θSup at Initial contact				
	Normal	11.1 (4.2)	11.5 (7.7–13.8)	0.57
	Slow	10.2 (3.7)	10.4 (7.6–12.9)	0.92
	Fast	12.0 (4.3)	12.6 (8.2–14.5)	0.71
	Medial	9.4 (3.5)	9.6 (6.5–11.9)	0.98
	Lateral	12.6 (4.1)	11.6 (10.4–14.6)	0.20
Maximum θTdown				
	Normal	61.9 (5.1)	64.2 (56.5–65.8)	0.06
	Slow	56.0 (5.1)	56.6 (52.2–59.4)	1.00
	Fast	68.5 (4.0)	68.7 (66.0–71.8)	0.29
	Medial	61.4 (3.9)	62.2 (58.5–64.2)	0.75
	Lateral	60.9 (3.8)	58.9 (58.4–63.9)	0.09
Maximum θPro				
	Normal	14.2 (7.7)	11.3 (9.1–18.8)	0.33
	Slow	11.2 (4.7)	9.2 (8.4–13.2)	0.02 *
	Fast	17.1 (9.5)	15.3 (9.8–19.1)	0.04 *
	Medial	12.5 (6.1)	11.4 (8.7–14.2)	0.21
	Lateral	13.5 (7.9)	10.8 (7.7–18.8)	0.52
Timing of maximum θTdown				
	Normal	70.1 (2.0)	70 (68–71.5)	0.03 *
	Slow	71.3 (1.4)	71 (70–72.5)	0.32
	Fast	67.9 (1.8)	67 (67–69.5)	0.29
	Medial	69.7 (1.8)	69 (68.5–70.5)	0.10
	Lateral	69.8 (1.4)	70 (69–71)	0.45
Timing of maximum θPro				
	Normal	74.7 (4.9)	72 (71–76)	<0.01 *
	Slow	76.8 (3.4)	77 (74.5–78)	0.31
	Fast	71.9 (3.9)	71 (69–73)	0.03 *
	Medial	75.8 (4.1)	74 (73–77)	0.03 *
	Lateral	75.3 (4.7)	75 (72–76.5)	0.12

SD: standard deviation. IQR: interquartile range. *: $p < 0.05$.

Written informed consent was obtained from all the participants. This study was approved by the Institutional Ethics Review Committee (PAZ21-31; 27 October 2021), and measurements were obtained with the participants' free and voluntary consent.

3. Results

The average angular changes during the five gait cycles are shown in Figure 3. And the data of each participant are shown in Table S1.

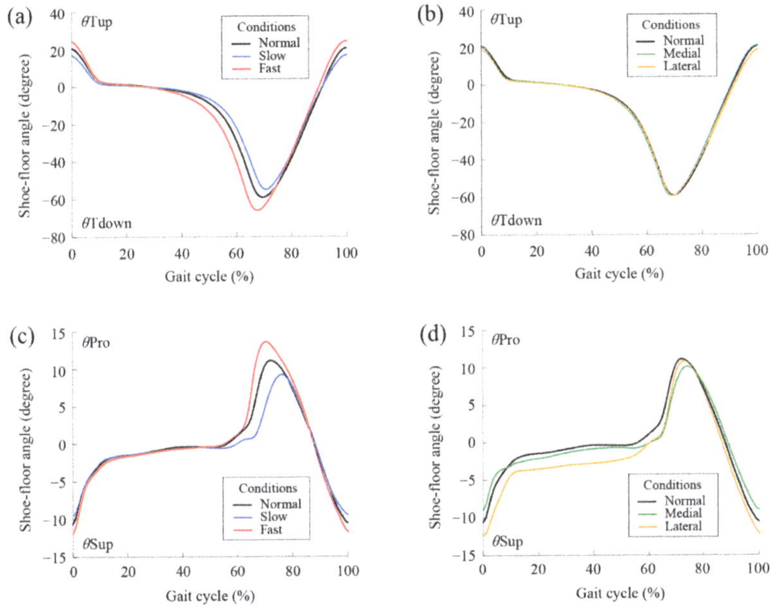

Figure 3. Angle changes of each condition are shown. Each waveform was calculated by averaging the angle change over five gait cycles. Angle change among speed conditions in the sagittal plane (**a**) and frontal plane (**c**). Angle change among sole adjustment conditions in the sagittal plane (**b**) and frontal plane (**d**). θTup indicates the angle in the dorsiflexion direction, and θTdown indicates the angle in the plantarflexion direction with respect to the floor. (**a**,**b**) are shown with θTup as positive. θPro indicates the angle in the eversion direction, and θSup indicates the angle in the inversion direction with respect to the floor. (**c**,**d**) are shown with θPro as positive.

3.1. Change in Magnitude of SFA with Gait Speed

Figure 4 shows the changes in the magnitude of the SFA with changes in speed. In the sagittal plane, the median θTup values at initial contact were 18.8°, 23.3°, and 25.4° in the Slow (3 km/h), Normal (4 km/h), and Fast (5 km/h) speed, respectively, and a significant effect of gait speed was found ($p < 0.01$). Significant differences were observed between the Slow and Normal speed ($p = 0.01$) and the Slow and Fast speed ($p < 0.01$); however, no significant difference between the Normal and Fast speed ($p = 0.05$) was noted.

The maximum θTdown values were 56.6°, 64.2°, and 68.7° under the Slow, Normal, and Fast speed, respectively, indicating a significant increase at higher speed ($p < 0.01$). Significant differences were observed between all speeds (Slow-Normal, $p = 0.02$; Normal-Fast, $p = 0.02$; Slow-Fast, $p < 0.01$).

In the frontal plane, the θSup values at initial contact were 10.4°, 11.5°, and 12.6° for the Slow, Normal, and Fast speed, respectively, and a significant effect of gait speed was demonstrated ($p = 0.04$). However, a statistically significant difference was observed only between the Slow and Fast speed ($p = 0.03$); there was no significant difference between the

Slow and Normal ($p = 0.60$) or Normal and Fast speed ($p = 0.60$). Finally, the maximum θPro values were 9.2°, 11.3°, and 15.3° for the Slow, Normal, and Fast speed, respectively, and a significant effect of gait speed was found ($p = 0.02$). A statistically significant difference was observed only between the Slow and Fast speed ($p = 0.02$), and no significant difference was found between the Slow and Normal ($p = 0.30$) and Normal and Fast speed ($p = 0.82$).

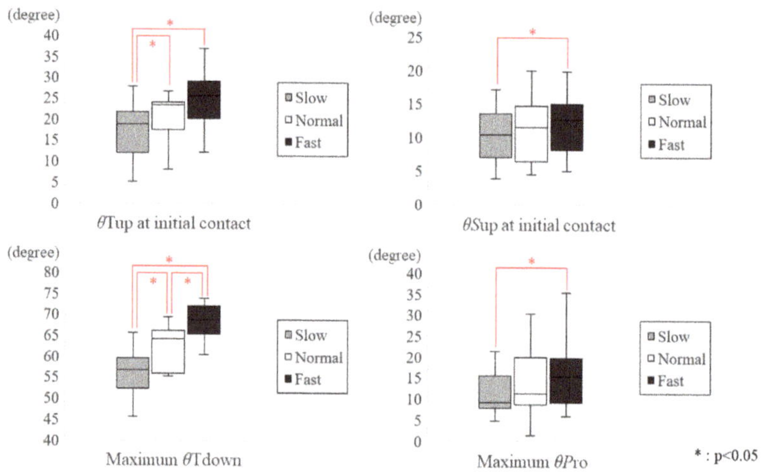

Figure 4. Change in shoe–floor angle with gait speed. *: $p < 0.05$.

3.2. Change in Magnitude of SFA with Sole Adjustment

The changes in the magnitude of the SFA due to the lift of one side of the sole are shown in Figure 5. A significant difference was observed only in the θSup at the initial contact ($p < 0.01$). Specifically, the increase in the θSup values were 11.5°, 9.6°, and 11.6° for the Normal, Medial, and Lateral conditions, respectively. Significant differences were observed only between the Medial and Lateral conditions ($p < 0.01$); however, there were no significant differences between the Normal and Medial ($p = 0.82$) or Normal and Lateral conditions ($p = 0.08$).

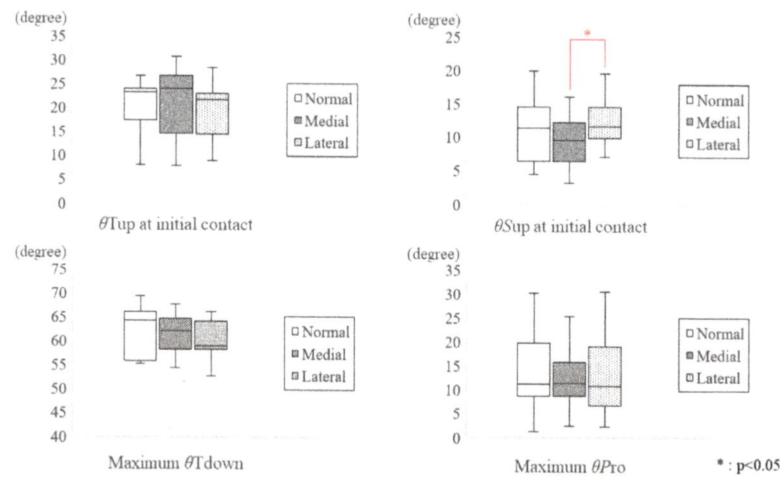

Figure 5. Change in shoe–floor angle with sole adjustment condition. *: $p < 0.05$.

3.3. *Change in Timing of SFA with Gait Speed and Sole Adjustment*

The results for the timing of the SFA are listed in Table 2. The timing is expressed as a percentage of the gait cycle with the gait cycle set as 100% (%gait cycle; %GC). The timing of the maximum θTdown were 71% GC, 70% GC, and 67% GC for the Slow, Normal, and Fast speed, respectively, and the timing significantly changed with gait speed ($p < 0.01$). Significant differences were observed between the Normal and Fast speed ($p < 0.01$), and between the Slow and Fast speed ($p < 0.01$).

Table 2. Change in timing of SFA with gait speed and sole adjustment.

% Gait Cycle (% GC)	Normal	Slow	Fast	Medial	Lateral
Toe down	70	71	67 *	69	70
Pronation	72	77	71 *	74	75

The data are expressed as 100% for each gait cycle (% gait cycle; % GC). The timing of the maximum toe-down and pronation were evaluated as timing of SFA. *: Significant difference compared with Normal.

The timing of maximum θPro were 77% GC, 72% GC, and 71% GC for the Slow, Normal, and Fast speed, respectively, and the timing significantly changed as speed increased ($p < 0.01$). Significant differences were observed between the Normal and Fast speed ($p = 0.02$) and between the Slow and Fast speed ($p < 0.01$).

No significant difference was observed in the effect of the sole adjustment on the timing of maximum θTdown and θPro ($p = 0.35$ and 0.19, respectively).

4. Discussion

The results of this study showed that a significant effect of gait speed was confirmed for several parameters relating to the magnitude of the SFA. Changes of sole adjustment affected only the supination angle at initial contact.

As shown in Figure 3, the shoe was grounded in the toe-up position, and the SFA changed to a toe-down position during the late stance. Additionally, the shoe was grounded in the supination position, in the loading response, it was in the intermediate position owing to shoe stiffness, and finally, it exhibited a maximum pronation position with the toe-off. Huang et al. [24] investigated the shoe–floor angle using a method similar to our study. The average waveform shown in this previous study is consistent with the waveform shown in this study (Figure 3), and the range of motion also shows similar values between both figures. Therefore, it is considered that the measurement used in this study is appropriate for indicating the shoe–floor angle.

The timing of the SFA demonstrated that the maximum pronation and toe-down occurred simultaneously during the toe-off in the late stance phase. In this phase, the center of pressure may move medially and forward from the heel because of the weight transfer to the opposite lower limb. Hence, it is suggested that the shoe pronation in the late stance phase is important for moving the center of foot pressure medially and guiding it to the thenar region during toe-off.

For the angle in the sagittal plane, both the toe-up and toe-down angles significantly increased with increasing gait speed. A previous study showed that dorsiflexion at the loading response, plantar flexion at the toe-off, and pronation and supination angles changed according to gait speed in the barefoot condition [20]. The results of this study showed similar changes in the SFA. These changes in angle may have occurred because of the increase in stride length. An increase in stride length may lead to an increase in the plantar flexion moment related to an increase in the angle of the toe-down. In a previous study, the foot angle in the late stance was related to forward propulsion [25]. An increase in the plantar flexion moment and forward propulsion may assist the swinging of the lower limb, resulting in a difference in the toe-up angle. A previous study showed that the foot angle in the sagittal plane was more strongly related to gait speed than to gait efficiency [14], which supports the results of this study. These results indicate that increased gait speed may expand the range of motion in the sagittal plane during fast walking. It is suggested that the contact area is reduced at initial contact and the toe-off phase. Therefore, higher balance ability and shoe flexibility may be needed for stability during fast walking. Furthermore, the results of this study suggested that shoes may contact the floor surface at the upper (proximal) positions of the toe and heel. When designing shoes for sports, it is important to adjust the flexibility of the forefoot and heel areas, and the choice of materials for the contact area with the floor.

Previous studies have reported changes in gait caused by sole adjustments in barefoot conditions. A previous study reported a reduction in the supination angle caused by a medial wedge insole [26]. Similarly, in this study, the supination angle tended to decrease in the Medial condition and slightly increase in the Lateral condition compared to that in the Normal condition, and a significant difference was found between the Medial and Lateral conditions. These results suggest that a change in the supination angle occurred to maintain a constant angle between the foot sole and floor surfaces. A previous study also showed the compensatory change to adapt to road conditions and shoe materials to prevent falls [27]. This result also suggested that the medial wedge insole may cause the displacement of the center of pressure toward a lateral position. The lateral area of the shoes needs to be robust and flexible. Furthermore, the results of this study showed that the lift of the lateral side of the sole increased the supination angle, suggesting that clinicians and designers should consider the possibility of an increased risk of inversion sprains when adjusting the shape of the sole surface.

This study has some limitations. First, the age range of the participants was narrow. The results may differ for older people and children because of differences in gait patterns compared to that for adults. However, these results need to be verified in a variety of participants, including patients and athletes. Second, the study was conducted in a laboratory setting, and the measurements were performed on a treadmill. A previous study showed a difference in gait between overground and treadmill walking [22]; therefore, future studies are required to investigate overground walking under outdoor conditions. Third, the relationship between the foot–floor angle and shoe–floor angle was not demonstrated. It is not clear to what extent a sole adjustment induces pronation and supination of the foot. To solve this problem, it is necessary to measure foot motion inside shoes during gait; however, this is currently difficult owing to the limitations of the measurement equipment. In the present study, the change in shoe movement was due to the gait conditions, but the reason of the results was unclear. In future studies, it will be necessary to consider a setting that allows simultaneous measurement of foot and shoe movements. Finally, for clinical applications, it is necessary to clarify the relationship among data measured using shoe-type sensors, motor function, and functional disorders. In the future, it will be necessary to clarify the correlation between the shoe–floor angle and the parameters used in medical and sports settings.

Shoe-type sensors do not require procedures such as attaching and fixing an accelerometer, allowing for simple measurements. Accelerometers are widely used in laboratory settings; however, in clinical settings, it is important to simplify the measurements to provide immediate feedback. The results of this study clarified the changes in the shoe–floor angle, and we believe that it has the potential for use in several clinical settings. A previous study showed that this sensor could be attached to commercially available shoes and to a patient's own shoes [19]. Thus, it may be possible to use this sensor repeatedly to adjust trial products and to assess gait function. In particular, the results of this study indicate that the contact area may differ depending on the walking speed and sole adjustment, which is an important issue for shoemakers to consider when maintaining the robustness and flexibility of shoes.

5. Conclusions

The results of this study demonstrated the SFA in response to changes in gait speed and showed that compensatory changes occurred in response to changes in the sole adjustment.

These results suggest that the magnitude of the SFA measured using ORPHE change according to gait speed, and the indices in the frontal plane can be indicators that reflect adaptive behavior. Shoe designers need to consider the change in the SFA to design shoes to prevent injuries and ensure durability. Thus, designing shoes according to the walking conditions is needed. Further studies are required to verify the SFA during various movements and activities of daily living. It is important to note that the ORPHE has potential as an assessment tool in gait analysis, shoemaking, and adjustment.

Supplementary Materials: The following supporting information can be downloaded at: https://www.mdpi.com/article/10.3390/biomechanics4040042/s1, Table S1: The data of each participant.

Author Contributions: Conceptualization, Y.H.; methodology, Y.H. and T.N.; validation, Y.H. and R.G.; formal analysis, Y.H. and T.N.; investigation, Y.H. and T.N.; resources, Y.H.; data curation, Y.H. and T.N.; writing—original draft preparation, Y.H.; writing—review and editing, Y.H. and R.G.; visualization, Y.H.; supervision, R.G.; project administration, Y.H.; funding acquisition, Y.H. All authors have read and agreed to the published version of the manuscript.

Funding: This research received no external funding.

Institutional Review Board Statement: The study was conducted in accordance with the Declaration of Helsinki and approved by the Ethics Review Committee of Gunma Paz University (PAZ21-31; 27 October 2021).

Informed Consent Statement: Informed consent was obtained from all subjects involved in the study.

Data Availability Statement: The data presented in this study are available in article and Supplementary Material.

Acknowledgments: The authors would like to thank Naoki Otsuka for technical support in the data collection and analysis process.

Conflicts of Interest: The authors declare no conflicts of interest.

References

1. Franklin, S.; Grey, M.J.; Heneghan, N.; Bowen, L.; Li, F.X. Barefoot vs common footwear: A systematic review of the kinematic, kinetic and muscle activity differences during walking. *Gait Posture* **2015**, *42*, 230–239. [CrossRef] [PubMed]
2. Keenan, G.S.; Franz, J.R.; Dicharry, J.; Della Croce, U.; Kerrigan, D.C. Lower limb joint kinetics in walking: The role of industry recommended footwear. *Gait Posture* **2011**, *33*, 350–355. [CrossRef] [PubMed]
3. Lythgo, N.; Wilson, C.; Galea, M. Basic gait and symmetry measures for primary school-aged children and young adults whilst walking barefoot and with shoes. *Gait Posture* **2009**, *30*, 502–506. [CrossRef] [PubMed]
4. Zhang, X.; Paquette, M.R.; Zhang, S. A comparison of gait biomechanics of flip-flops, sandals, barefoot and shoes. *J. Foot Ankle Res.* **2013**, *6*, 45. [CrossRef]
5. Oeffinger, D.; Brauch, B.; Cranfill, S.; Hisle, C.; Wynn, C.; Hicks, R.; Augsburger, S. Comparison of gait with and without shoes in children. *Gait Posture* **1999**, *9*, 95–100. [CrossRef]
6. Sacco, I.C.; Akashi, P.M.; Hennig, E.M. A comparison of lower limb EMG and ground reaction forces between barefoot and shod gait in participants with diabetic neuropathic and healthy controls. *BMC Musculoskelet. Disord.* **2010**, *11*, 24. [CrossRef]
7. Vienne, A.; Barrois, R.P.; Buffat, S.; Ricard, D.; Vidal, P.P. Inertial Sensors to Assess Gait Quality in Patients with Neurological Disorders: A Systematic Review of Technical and Analytical Challenges. *Front. Psychol.* **2017**, *8*, 817. [CrossRef]
8. Menz, H.B.; Lord, S.R.; Fitzpatrick, R.C. Acceleration patterns of the head and pelvis when walking on level and irregular surfaces. *Gait Posture* **2003**, *18*, 35–46. [CrossRef]
9. Moe-Nilssen, R.; Helbostad, J.L. Interstride trunk acceleration variability but not step width variability can differentiate between fit and frail older adults. *Gait Posture* **2005**, *21*, 164–170. [CrossRef]
10. Smidt, G.L.; Arora, J.S.; Johnston, R.C. Accelerographic analysis of several types of walking. *Am. J. Phys. Med.* **1971**, *50*, 285–300.
11. Yentes, J.M.; Raffalt, P.C. Entropy Analysis in Gait Research: Methodological Considerations and Recommendations. *Ann. Biomed. Eng.* **2021**, *49*, 979–990. [CrossRef] [PubMed]
12. Hausdorff, J.M. Gait dynamics, fractals and falls: Finding meaning in the stride-to-stride fluctuations of human walking. *Hum. Mov. Sci.* **2007**, *26*, 555–589. [CrossRef] [PubMed]
13. Brégou Bourgeois, A.; Mariani, B.; Aminian, K.; Zambelli, P.Y.; Newman, C.J. Spatio-temporal gait analysis in children with cerebral palsy using, foot-worn inertial sensors. *Gait Posture* **2014**, *39*, 436–442. [CrossRef] [PubMed]

14. Wert, D.M.; Brach, J.; Perera, S.; VanSwearingen, J.M. Gait biomechanics, spatial and temporal characteristics, and the energy cost of walking in older adults with impaired mobility. *Phys. Ther.* **2010**, *90*, 977–985. [CrossRef] [PubMed]
15. Moyer, B.E.; Chambers, A.J.; Redfern, M.S.; Cham, R. Gait parameters as predictors of slip severity in younger and older adults. *Ergonomics* **2006**, *49*, 329–343. [CrossRef]
16. Baba, T.; Watanabe, M.; Ogihara, H.; Handa, S.; Sasamoto, K.; Okada, S.; Okuizumi, H.; Kimura, T. Validity of temporo-spatial characteristics of gait as an index for fall risk screening in community-dwelling older people. *J. Phys. Ther. Sci.* **2023**, *35*, 265–269. [CrossRef]
17. Menz, H.B.; Dufour, A.B.; Riskowski, J.L.; Hillstrom, H.J.; Hannan, M.T. Foot posture, foot function and low back pain: The Framingham Foot Study. *Rheumatology* **2013**, *52*, 2275–2282. [CrossRef]
18. Kuhman, D.J.; Paquette, M.R.; Peel, S.A.; Melcher, D.A. Comparison of ankle kinematics and ground reaction forces between prospectively injured and uninjured collegiate cross country runners. *Hum. Mov. Sci.* **2016**, *47*, 9–15. [CrossRef]
19. Uno, Y.; Ogasawara, I.; Konda, S.; Yoshida, N.; Otsuka, N.; Kikukawa, Y.; Tsujii, A.; Nakata, K. Validity of Spatio-Temporal Gait Parameters in Healthy Young Adults Using a Motion-Sensor-Based Gait Analysis System (ORPHE ANALYTICS) during Walking and Running. *Sensors* **2022**, *23*, 331. [CrossRef]
20. van Hoeve, S.; Leenstra, B.; Willems, P.; Poeze, M.; Meijer, K. The effect of age and speed on foot and ankle kinematics assessed using a 4-segment foot model. *Medicine* **2017**, *96*, e7907. [CrossRef]
21. Finley, F.R.; Cody, K.A. Locomotive characteristics of urban pedestrians. *Arch. Phys. Med. Rehabil.* **1970**, *51*, 423–426. [PubMed]
22. Jung, T.; Kim, Y.; Kelly, L.E.; Wagatsuma, M.; Jung, Y.; Abel, M.F. Comparison of Treadmill and Overground Walking in Children and Adolescents. *Percept. Mot. Ski.* **2021**, *128*, 988–1001. [CrossRef] [PubMed]
23. Madgwick, S.O.; Harrison, A.J.; Vaidyanathan, A. Estimation of IMU and MARG orientation using a gradient descent algorithm. In Proceedings of the 2011 IEEE International Conference on Rehabilitation Robotics, Zurich, Switzerland, 29 June–1 July 2011; Volume 2011, p. 5975346. [CrossRef]
24. Huang, C.; Nihey, F.; Ihara, K.; Fukushi, K.; Kajitani, H.; Nozaki, Y.; Nakahara, K. Healthcare Application of In-Shoe Motion Sensor for Older Adults: Frailty Assessment Using Foot Motion during Gait. *Sensors* **2023**, *23*, 5446. [CrossRef] [PubMed]
25. Ensink, C.J.; Hofstad, C.; Theunissen, T.; Keijsers, N.L.W. Assessment of Foot Strike Angle and Forward Propulsion with Wearable Sensors in People with Stroke. *Sensors* **2024**, *24*, 710. [CrossRef]
26. Costa, B.L.; Magalhães, F.A.; Araújo, V.L.; Richards, J.; Vieira, F.M.; Souza, T.R.; Trede, R. Is there a dose-response of medial wedge insoles on lower limb biomechanics in people with pronated feet during walking and running? *Gait Posture* **2021**, *90*, 190–196. [CrossRef]
27. Menant, J.C.; Steele, J.R.; Menz, H.B.; Munro, B.J.; Lord, S.R. Effects of walking surfaces and footwear on temporo-spatial gait parameters in young and older people. *Gait Posture* **2009**, *29*, 392–397. [CrossRef]

biomechanics

Article

Retention and Transfer of Fractal Gait Training

Logan J. Frame [1], Nikita A. Kuznetsov [2], Louisa D. Raisbeck [1] and Christopher K. Rhea [3,*]

[1] Department of Kinesiology, University of North Carolina Greensboro, Greensboro, NC 27412, USA; loganjohnframedpt@gmail.com (L.J.F.); ldraisbe@uncg.edu (L.D.R.)
[2] Department of Rehabilitation, Exercise & Nutrition Sciences, University of Cincinnati, Cincinnati, OH 45220, USA; kuznetna@ucmail.uc.edu
[3] Ellmer College of Health Sciences, Old Dominion University, Norfolk, VA 23529, USA
* Correspondence: crhea@odu.edu

Abstract: Background/Purpose: Fractal gait patterns have been shown to be modifiable, but the extent to which they are retained and transferred to new contexts is relatively unknown. This study aimed to close those gaps by enrolling participants (N = 23) in a seven-day fractal gait training program. Methods: Building on related work, the fractal gait training occurred on a treadmill over a 10-min period. Before and after the treadmill training, each participant walked for 10 min overground without the fractal stimulus used during training. The daily post-test was used to examine immediate retention and transfer of the fractal gait patterns from the treadmill to overground. The pre-tests in days 2–7 were used to examine the extent to which the fractal gait patterns from the preceding day were retained 24 h later. Inertial measurement units were used to measure stride time so a consistent measurement method could be employed in the treadmill and overground phases of the study. Results: Our results showed that multiple days of treadmill training led to elevated fractal patterns, indicating a positive training effect. However, the positive training effect observed on the treadmill did not transfer to overground walking. Conclusions: Collectively, the data show that fractal patterns in gait are modifiable across multiple days of training, but the transferability of these patterns to new contexts needs to be further explored.

Keywords: gait; locomotion; mobility; fractal; inertial sensors

1. Introduction

It was observed nearly three decades ago that the natural variation from stride-to-stride during steady-state gait is not random, but rather, exhibits a fractal pattern [1]. Follow-up work showed that task constraints [2–7], injury [8], aging across the lifespan [9,10], and disease [9,11,12] can alter the fractal patterns. Beyond a quantitative observation, it has been suggested that the nature of fractal patterns relate to gait adaptability [13] and fall risk [14], an assertion that links back to the original implementation of fractal pattern analysis in physiological systems [15,16]. The connection to adaptability has been strengthened with the recent observation that fractal patterns—in particular, pink noise—promote earlier coordination transitions in movement [17], reinforcing the postulate that fractal patterns afford flexible movement options that can be adopted based on the needs of the task.

With the acknowledgement that fractal gait patterns change after injury, aging, or disease, and the postulate that they reflect a person's ability to exhibit adaptive movement, it stands to reason that enhancing fractal gait patterns in clinical populations could lead to positive outcomes, such as a decline in fall risk [18]. One potential way to alter fractal patterns in movement is to provide a fractal stimulus to which participants can synchronize their movement. The theoretical framework for such an intervention was first articulated by Dubois using a mathematical model of strong and weak anticipations [19] and further expanded upon by West and colleagues with the complexity matching framework [20].

This phenomenon was first demonstrated empirically in human movement data utilizing a finger tapping paradigm [21]. In that study, Stephen and colleagues asked participants to tap a key on the keyboard in synchrony with each flash on the screen in front of them. The visual stimulus on the screen acted as a visual metronome, similar to the rhythmic auditory stimulation (RAS) paradigm commonly used in rehabilitation where an auditory metronome provides a cue to which a patient attempts to synchronize their movements [22,23]. However, there is one very important distinction differentiating the metronome used by Stephen and colleagues. Typical RAS paradigms utilize a consistent time interval between the "beeps", whereas the metronome used by Stephen and colleagues utilized a variable time interval between the stimulus presentation, such that the variation in time intervals exhibited a fractal structure. With this implementation, Stephen and colleagues created a fractal metronome and postulated that participants would be able to synchronize to the metronome, using the strong anticipation framework from Dubois as their rationale. Their hypothesis was supported, and they showed that the fractal structure of movement patterns could be influenced by a fractal metronome.

A logical extension of this work was to determine if similar synchronization behavior occurred in gait. Some early studies employed a fractal auditory metronome to modify fractal gait patterns [24,25]. Rhea et al. used a visual metronome to more closely replicate and extend Stephen et al.'s work to gait [26]. Their study showed that fractal gait patterns in young healthy adults could be either strengthened or weakened, depending on the structure of the fractal visual stimulus. The observation that fractal gait patterns can be modified using a fractal stimulus has since been replicated in various forms [27–33]. Moreover, it has been shown that the newly developed fractal gait patterns are retained—at least in the short-term—after training [24,27,34].

While fractal gait training appears promising, it should be noted that the retention characteristics of this type of training are relatively unknown. For example, 15 min of training led to strengthened fractal gait patterns in the post-test immediately after the training, but the strength of the patterns began to drop back down at the end of the retention period (albeit still elevated relative to the pre-training fractal levels) [34]. In patients with Parkinson's disease, it was shown that fractal patterns could be altered during training and 5 min afterwards, but more extended retention was not assessed [24]. In another study, four days of fractal gait training were shown to increase the strength of the fractal gait patterns, but retention was not assessed [35]. Thus, the effect of multiple days of fractal gait training on immediate and longer-term retention is unclear. Furthermore, it is unknown whether fractal gait training on a treadmill transfers to overground walking, which is important to know not only from a neuromotor control perspective, but from a practical perspective in order to help identify the efficacy of this type of training for use in rehabilitation settings. The incorporation of inertial measurement units (IMUs) in this line of research is important, as they afford the possibility of a consistent data collection method across treadmill and overground settings. Video-based motion capture is not a good candidate for this line of research since typically hundreds of strides are necessary to reliably calculate fractal gait patterns [36]. Such a volume is difficult to capture with video-based motion capture, especially in overground gait in large areas, making IMUs a stronger data collection candidate.

In summary, there is a long history of examining fractal gait patterns in the context of injury [8], aging across the lifespan [9,10], and disease [9,11,12]. What has been consistently shown is that fractal gait patterns change in these contexts. The extent to which these changes have any functional meaning is still an empirical question. Some have postulated that fractal patterns are related to fall risk and/or adaptability based on empirical evidence [13,14,18,37,38]. However, these observations lack large sample sizes with a variety of populations, which would provide stronger evidence for the connection between fractal gait patterns and adaptability. Nevertheless, the empirical observations showing that fractal gait patterns may be related to adaptability provide credence for the exploration of pathways to positively alter the fractal gait pattern. It is important to show this proof-of-concept first in healthy populations to avoid burdening clinical populations unnecessarily. Our study tackles this issue by exploring the extent to which fractal gait patterns in healthy young adults are modifiable over a 7-day training period.

The purpose of this study was to determine the extent to which training to synchronize stride timing to a fractal visual metronome while walking on a treadmill would affect the adopted stride time variability pattern during overground locomotion. Two hypotheses were tested: (1) participants would exhibit stronger coupling to the fractal visual metronome with practice and (2) immediate retention (i.e., directly after training) and longer-term retention (i.e., 24 h after training) would increase as a function of increased practice.

2. Materials and Methods

2.1. Participants

A convenience sample of healthy young adults (N = 23; 22.0 ± 2.5 yrs; 172.7 ± 11.1 cm) primarily consisting of students from the University of North Carolina at Greensboro (UNCG) were enrolled in the study. Potential volunteers who were notified of the opportunity to participate via flyers in public locations, presentations in their courses by a member of the study team, and via word of mouth. Participants self-reported to be free of neuromuscular injuries or pathologies that might cause abnormal biomechanics and had normal or corrected-to-normal vision. Study procedures were approved by the Institutional Review Board (IRB) at the University of North Carolina at Greensboro (IRB study number 15-0370).

2.2. Materials

Three IMUs from the APDM Mobility Lab Opal system (APDM, Inc., Portland, OR, USA) were attached to the participants with elastic straps: two on top of each foot (center on the front of the ankle) and one on the lumbar spine (centered at the base of the spine), identical to the ADPM sensor placement in [39]. Stride time was extracted from the acceleration and orientation time series recorded by the APDM sensors using APDM Mobility Lab software version 2, which has been shown to provide valid measurement of gait stride time [40–42]. Sensor data were collected at 128 Hz.

2.3. Procedures

Participants synchronized with a fractal visual metronome over seven consecutive sessions separated by 24 h (±30 min). Within each session they completed a pre-test, training, and post-test phases, each phase lasting 10 min. The pre-test phase consisted of overground walking at preferred speed around a 62 m gymnasium track. The training phase consisted of walking on a treadmill at individually determined preferred walking speed while synchronizing left and right leg steps with the appearance of the corresponding foot on display positioned on the treadmill panel directly in front of them (Figure 1; Apple iPad, Cupertino, CA, USA). The timing of appearance of the right footprint was determined by a strongly persistent fractal time series, as characterized by detrended fluctuation analysis alpha (DFA α) of 0.98 (Figure 2), consistent with our previous research on fractal gait training [26,34]. Fractal time series contained 500 data points bounded within 1.00–1.35 s (1.17 ± 0.07 s). Left footprints appeared halfway through each right foot phase. The post-test phase was identical to the pre-test. Participants were asked to maintain consistent walking speed throughout the sessions.

Preferred treadmill walking speed was determined for each participant in the first session using the following procedure: speed of the treadmill was increased from 0.5 m/s in increments of 0.05 m/s until participant verbally indicated that a given speed was their normal walking speed. The second estimate was started from the speed that was too fast (2.0 m/s) and incrementally slowed until they verbally indicated that the speed was their normal pace. If the two speeds were within 10% of each other, the average of the two speeds was set as the preferred walking speed. If there was more than 10% difference in the speeds, participant completed the procedure again. All subjects self-selected a walking speed within less than two attempts. This speed was used for all consequent treadmill training sessions.

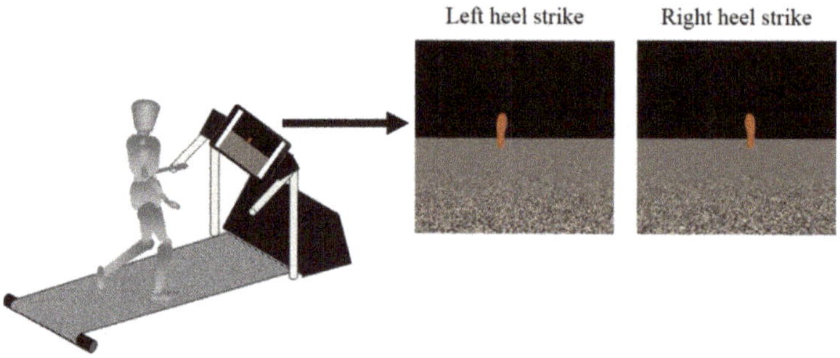

Figure 1. Schematic of the experimental setup. An iPad was placed on the treadmill that presented a video of a left and right footprint that flashed at prescribed timing intervals exhibiting a fractal structure. Participants were instructed to be at heel strike when the corresponding foot appeared.

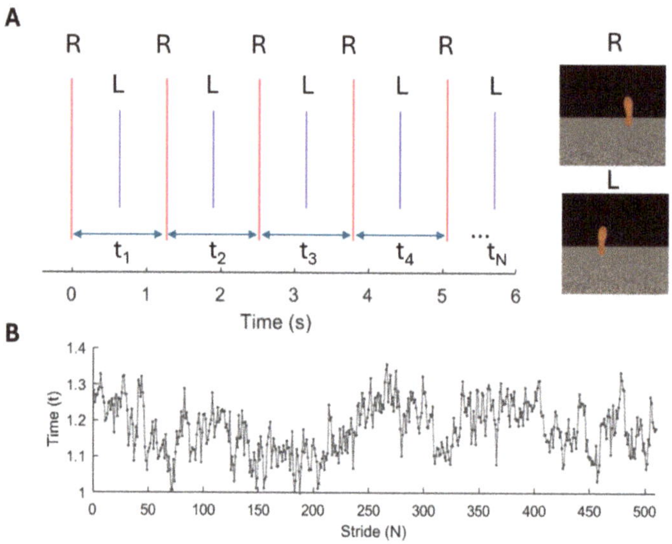

Figure 2. Participants synchronized their heel strikes to the visual display presented to them. The time between the appearance of the right heel strikes (panel (**A**)) was determined by the persistent fractal time series presented in panel (**B**). Left foot strikes appeared at half-time between consecutive right heel strikes. "R" represents when the right heel strike was prescribed to occur, whereas "L" represents when the left heel strike was prescribed to occur.

2.4. Data Analysis

Adaptive Fractal Analysis (AFA) [43–45] was used to characterize serial correlations in the stride interval time series produced by right foot heel strikes. This method is more accurate than the DFA for short signals [46] and when signals have nonlinear trends [43,45]. A full tutorial on AFA has been previously published [47]. Briefly, AFA uses the following steps: (1) integrate the time series; (2) generate a globally smooth trend signal to the integrated time series by patching together local polynomial fits to the data at a time scale determined by the selected window size (w); (3) calculate variance of the data around the polynomial trend (termed fluctuation function, F(w)); (4) examine the relationship between the F(w) as function of w on a log-log plot. The slope of the linear relationship between $\log_2$ (F(w)) and $\log_2$ (w) provides a characterization of the serial correlations in the time series: $\alpha = 0.5$ indicates random noise, $0 < \alpha < 0.5$—anti-persistent dynamics, $0.5 < \alpha < 1$—persistent dynamics. When α values are greater than 1, this indicates a different class of signals: Brownian noise processes ($1 < \alpha < 1.5$—under-diffusive random walk, $\alpha = 1.5$—random walk, $1.5 < \alpha < 2$—hyper-diffusive random walk.

We estimated the scaling exponent α within two regions similarly to Marmelat et al. [31]: a short-term region (αST) spanning window sizes from 5 to 9 strides and long-term region (αLT) spanning from 13 to 183 strides. The rationale for choosing these regions was that participants have been reported to synchronize to unpredictable fractal signals only on long-term scales (termed "complexity matching"), even without being able to synchronize on the short stride-by-stride scale. The increase in global coupling over practice is illustrated for a single subject in Figure 3. All analyses were performed in Matlab 2017a (Mathworks, Natick, MA, USA).

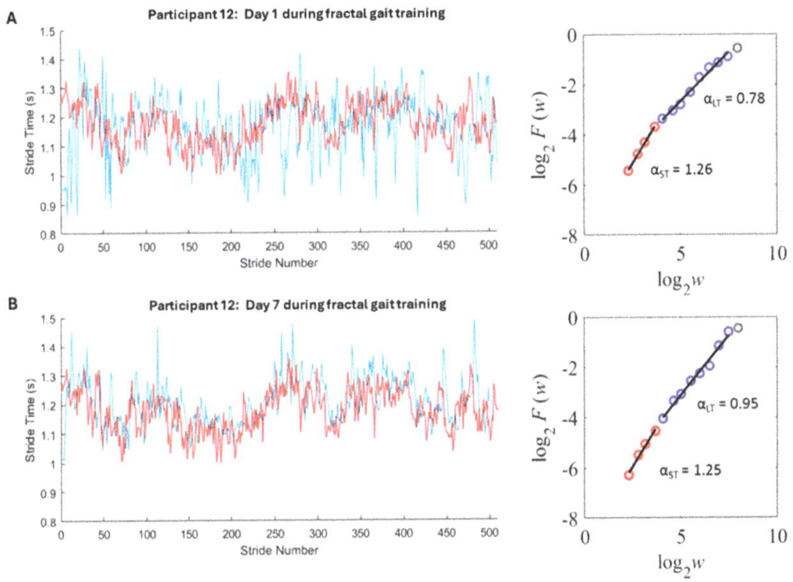

Figure 3. Illustration of increased global coupling during fractal gait training on the treadmill between the right heel stride time (blue) and the fractal metronome (red) as indexed by the long-term scaling exponent (αLT) from day 1 (panel (**A**)) to day 7 (panel (**B**)). The short-term scaling exponent (αST, red circles in log-log plot) stayed consistent across the seven days of fractal gait training, whereas the long-term scaling exponent (αLT, blue circles in log-log plot) shifted toward more persistent behavior.

2.5. Statistical Analysis

Linear mixed models (LMM) were used to evaluate the effects of metronome training on gait performance. An LLM approach was better suited for our study design relative to a typical repeated measures analysis of variance (ANOVA) approach for several reasons. First, an ANOVA assumes data normality, homogeneity, and sphericity, all of which can be particularly challenging with longitudinal datasets. The LLM approach allows for different variances across levels or groups, affording greater flexibility to model the data. Regarding sphericity, an ANOVA assumes that variance is equal between the combinations of levels or groups, whereas LLM does not have this assumption and can model the covariance in the data. Lastly, LLM can model individual difference trajectories over time, which is important for longitudinal study designs. The LLM analysis for our study design led to the modeling of an estimated marginal mean at each session (day) and condition (pre-test, training, post-test), which afford the comparison of across the longitudinal analysis. Collectively, the flexibility and robustness afforded by an LLM approach was optimal for our longitudinal study design.

To address Hypothesis 1, we examined change of αST and αLT during the synchronization phase over 7 sessions. The null model was compared to the model with the main effect of session to evaluate statistical significance (alpha = 0.05). To address Hypothesis 2, we used LMM with Condition (pre- and post-test) and Session (1 through 7) on αST and αLT. The main effects and the interaction were evaluated by comparing LMM with and without these effects. All statistical analyses were performed in R version 3.3.3 using the lme4 toolbox.

3. Results

3.1. *Hypothesis 1: Participants Exhibit Stronger Coupling to the Fractal Visual Metronome with Practice*

Participants successfully modified their gait dynamics to synchronize with the fractal metronome, as both αST and αLT were greater during the synchronization phase compared to pre- and post-test (Figure 4). However, upon visual inspection of individual data, we identified four participants (S1, S5, S8, and S11) who were not able to synchronize to the metronome—they did not match the period of the metronome and thus did not follow task instruction. Their data were excluded from further analyses.

Local coupling between right leg stride time and fractal visual metronome during synchronization phase did not change with practice as indicated by constant αST over the seven practice sessions ($p > 0.05$; Figure 4A). Values of αST were greater than 1, indicating strongly persistent under-diffusive Brownian motion dynamics of stride time on short-term time scales. Global coupling of stride time to the metronome, however, became stronger with practice as indicated by an increase in the αLT toward to the exponent prescribed by the metronome ($p < 0.001$; Figure 4B).

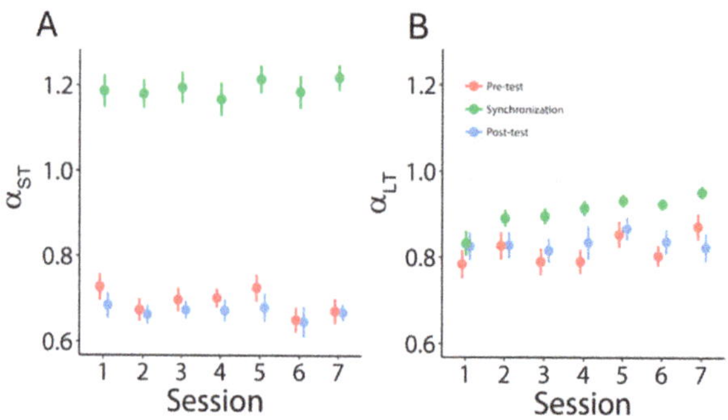

Figure 4. Illustration of increased global coupling of right heel stride time during treadmill trials (blue) to the fractal metronome (red) as indexed by the long-term scaling exponent from Session 1 to Session 7. Error bars represent standard error. The short-term scaling exponent (αST) is presented in panel (**A**) and the long-term scaling exponent (αLT) is presented in panel (**B**).

3.2. *Hypothesis 2: Immediate Retention (i.e., Directly After Training) and Longer-Term Retention (i.e., 24 h After Training) Would Increase as a Function of Increased Practice*

There was an immediate effect of metronome synchronization treadmill training on short-term dynamics of stride time during overground walking, such that αST became smaller in the post-test phase compared to the pre-test phase ($p < 0.001$; Figure 4A). This suggests that the short-term dynamics of stride time became more random immediately after metronome training. However, metronome training on the treadmill did not alter the long-term scaling exponent during overground locomotion, as there was no statistical difference between the pre- and post-test long-term scaling exponent in all testing sessions (Figure 4B).

4. Discussion

The purpose of this study was to determine the extent to which training to synchronize stride timing to a fractal visual metronome while walking on a treadmill would affect the adopted stride time fractal pattern during overground gait. To our knowledge, this is the first study to explore fractal gait training across multiple days that also included a transfer test to overground. Our most impactful finding was that multiple days of treadmill training led to elevated fractal patterns in the long-term region, indicating a positive training effect. However, the effect of positive training that was observed on the treadmill did not transfer to overground walking.

Our first hypothesis was partially supported with the observation that both the local coupling (αST) and global coupling (αLT) were elevated during the treadmill training. However, the local coupling did not increase across days, whereas the global coupling exhibited a relatively linear increase across days. These data provide support for the complexity matching framework, such that the one system attempts to synchronize with the global dynamics exhibited by a connected system [20]. Our data replicate and extend the findings from Uchitomi et al. where they showed a positive training effect from synchronizing to a fractal metronome over four training sessions on consecutive days [35]. Our seven consecutive day study design shows that the linear increase in fractal patterns can continue beyond the initial four-day period tested by Uchitomi and colleagues. Furthermore, our results align with a study showing that higher dosage of training positively effects the enhancement of movement patterns [48].

An important extension of our work is the separation of short- and long-term regions in the AFA analysis, consistent with the approach taken by Marmelat et al. [31], whereas Uchitomi et al. utilized a single region for their DFA analysis. By utilizing short- and long-term regions, we were able to identify differentiating effects of the training. That is, the short-term region was elevated—indicating a response to the fractal training—but the response was not strengthened (or weakened) across the training days. Conversely, the long-term region showed a consistent increase across days, highlighting a dose response training effect. It is important to note that Uchitomi et al. used DFA to characterize fractal gait patterns in their study. While DFA has a long history of being utilized in this context [1,2,49], its challenges with short signals [46] and nonlinear trends [43,45] have been identified. Our use of AFA, an analysis with a similar mathematical foundation as DFA, prohibits a direct comparison to Uchitomi et al., but does provide a deeper dive into effects from fractal gait training.

For the second hypothesis, the results show that there was no immediate retention or 24-h retention of the fractal gait behavior adopted during the training phase. This is inconsistent with previous work that showed immediate retention when the participant stayed on the treadmill during the post-testing phase [34]. The current study was designed to extend the previous study by examining retention during a transfer test. Thus, immediate retention was tested by having the participants walk overground directly after the treadmill training, whereas longer-term retention was assessed 24 h later via the overground walking pre-test that preceded that day's fractal gait training on the treadmill. The lack of retention and transfer on the same day suggests context dependence with respect to the training. The gait behavior learned during the training phase on the treadmill was not extended to overground walking. This was not completely surprising, as previous work has shown that spatiotemporal gait parameters can differ between treadmill and overground walking, especially with respect to gait variability [50,51]. The reasons for these differences may be contextual. For example, a treadmill provides a smaller ground reaction force than overground gait [52]. A smaller ground reaction force may lead to less salient feedback about foot position during the gait cycle, so a task learned on the treadmill may not be transferred to overground gait due to different ground reaction force profiles. Furthermore, the continuous belt of the treadmill propagates a movement that is outside of the body. The treadmill will not stop even if the subject does. This continuous movement may cause a sensory difference between the two conditions. The self-selected gait speed may have been different enough during overground gait to contaminate any learning that occurred on the treadmill. Previous work has shown that DFA α of stride time intervals changes with different gait speeds [4], so not precisely controlling for different walking speeds on the treadmill and overground may have influenced our results.

A unique contribution made by our study design was to explore the transferability of fractal patterns learned during a training session, which had not yet been empirically explored. Building on the growing literature of fractal gait training [24–33] and retention of the newly developed fractal patterns [24,27,34], exploring the extent to which treadmill training (which is conducive to clinical/rehabilitation programs) is transferred to overground gait (which aligns with activities of daily living) was a logical next step. Our data shows that the training paradigm can lead to stronger fractal patterns, but those patterns may be contextual and/or environmentally dependent. From a practical perspective, our data suggest that fractal gait training on a treadmill may not a plausible mechanism for altering fractal gait characteristics overground. Future studies should extend the previous fractal gait training work in overground contexts by adding a multi-day training component to explore the retention of the newly developed gait patterns in short- and longer-term durations. Furthermore, future studies should empirically test the extent to which fractal gait patterns relate to adaptability, such as recovery from a trip to avoid a fall. Such information would more firmly ground fractal gait training as a plausible addition to clinical rehabilitation. The continued adoption of IMUs in this context will afford the testing of fractal gait training across a variety of overground environments.

5. Conclusions

This study showed a dose response effect for fractal gait training across a seven-day period with respect to the global coupling to the metronome, the first observation of its kind. Although a stronger fractal pattern was observed during training across the days, transfer of the new pattern to overground gait was not observed. The innovation of our study design was the deployment of a fractal gait training paradigm over a seven-day period, with the limitation that only young healthy adults were enrolled. Future studies should focus on deploying fractal gait training in an overground environment, enrolling a wider spectrum of participants to measure generalizability, and develop study designs to empirically test the relationship between fractal gait patterns and adaptability. Collectively, the data show that fractal patterns in gait are modifiable, but the transferability of these patterns to new contexts needs to be further explored.

Author Contributions: Conceptualization, L.J.F. and C.K.R.; methodology, L.J.F., N.A.K., L.D.R. and C.K.R.; software, L.J.F., N.A.K. and C.K.R.; validation, L.J.F.; formal analysis, L.J.F. and N.A.K.; investigation, L.J.F.; resources, L.D.R. and C.K.R.; data curation, L.J.F.; writing—original draft preparation, L.J.F.; writing—review and editing, L.J.F., N.A.K., L.D.R. and C.K.R.; visualization, L.J.F. and C.K.R.; supervision, C.K.R.; project administration, L.J.F. All authors have read and agreed to the published version of the manuscript.

Funding: This research received no external funding.

Institutional Review Board Statement: The study was conducted in accordance with the Declaration of Helsinki and approved by the Institutional Review Board of the University of North Carolina at Greensboro (IRB study number 15-0370, approval date: 28 August 2015).

Informed Consent Statement: Informed consent was obtained from all subjects involved in the study.

Data Availability Statement: Data will be made available upon reasonable request to the corresponding author.

Conflicts of Interest: The authors declare no conflicts of interest.

References

1. Hausdorff, J.M.; Peng, C.K.; Ladin, Z.; Wei, J.Y.; Goldberger, A.L. Is walking a random walk? Evidence for long-range correlations in stride interval of human gait. *J. Appl. Physiol.* **1995**, *78*, 349–358. [CrossRef] [PubMed]
2. Hausdorff, J.M.; Purdon, P.L.; Peng, C.K.; Ladin, Z.; Wei, J.Y.; Goldberger, A.L. Fractal dynamics of human gait: Stability of long-range correlations in stride interval fluctuations. *J. Appl. Physiol.* **1996**, *80*, 1448–1457. [CrossRef] [PubMed]
3. Jordan, K.; Challis, J.H.; Newell, K. Long range correlations in the stride interval of running. *Gait Posture* **2006**, *24*, 120–125. [CrossRef] [PubMed]
4. Jordan, K.; Challis, J.H.; Newell, K. Walking speed influences on gait cycle variability. *Gait Posture* **2007**, *26*, 128–134. [CrossRef] [PubMed]
5. Jordan, K.; Challis, J.H.; Newell, K. Speed influences on the scaling behavior of gait cycle fluctuations during treadmill running. *Hum. Mov. Sci.* **2007**, *26*, 87–102. [CrossRef]
6. Armitano-Lago, C.; Pietrosimone, B.; Davis-Wilson, H.C.; Evans-Pickett, A.; Franz, J.R.; Blackburn, T.; Kiefer, A.W. Biofeedback augmenting lower limb loading alters the underlying temporal structure of gait following anterior cruciate ligament reconstruction. *Hum. Mov. Sci.* **2020**, *73*, 102685. [CrossRef]
7. Terrier, P.; Turner, V.; Schutz, Y. GPS analysis of human locomotion: Further evidence for long-range correlations in stride-to-stride fluctuations of gait parameters. *Hum. Mov. Sci.* **2005**, *24*, 97–115. [CrossRef]
8. Meardon, S.A.; Hamill, J.; Derrick, T.R. Running injury and stride time variability over a prolonged run. *Gait Posture* **2011**, *33*, 36–40. [CrossRef]
9. Hausdorff, J.M.; Mitchell, S.L.; Firtion, R.; Peng, C.K.; Cudkowicz, M.E.; Wei, J.Y.; Goldberger, A.L. Altered fractal dynamics of gait: Reduced stride-interval correlations with aging and Huntington's disease. *J. Appl. Physiol.* **1997**, *82*, 262–269. [CrossRef]
10. Hausdorff, J.M.; Zemany, L.; Peng, C.K.; Goldberger, A.L. Maturation of gait dynamics: Stride-to-stride variability and its temporal organization in children. *J. Appl. Physiol.* **1999**, *86*, 1040–1047. [CrossRef]
11. Hausdorff, J.M.; Lertratanakul, A.; Cudkowicz, M.E.; Peterson, A.L.; Kaliton, D.; Goldberger, A.L. Dynamic markers of altered gait rhythm in amyotrophic lateral sclerosis. *J. Appl. Physiol.* **2000**, *88*, 2045–2053. [CrossRef] [PubMed]
12. Moon, Y.; Sung, J.; An, R.; Hernandez, M.E.; Sosnoff, J.J. Gait variability in people with neurological disorders: A systematic review and meta-analysis. *Hum. Mov. Sci.* **2016**, *47*, 197–208. [CrossRef] [PubMed]
13. Ducharme, S.W.; Liddy, J.J.; Haddad, J.M.; Busa, M.A.; Claxton, L.J.; van Emmerik, R.E. Association between stride time fractality and gait adaptability during unperturbed and asymmetric walking. *Hum. Mov. Sci.* **2018**, *58*, 248–259. [CrossRef] [PubMed]
14. Hausdorff, J.M. Gait dynamics, fractals and falls: Finding meaning in the stride-to-stride fluctuations of human walking. *Hum. Mov. Sci.* **2007**, *26*, 555–589. [CrossRef]
15. Lipsitz, L.A.; Goldberger, A.L. Loss of 'complexity' and aging: Potential applications of fractals and chaos theory to senescence. *JAMA J. Am. Med. Assoc.* **1992**, *267*, 1806–1809. [CrossRef]
16. Lipsitz, L.A. Dynamics of stability: The physiologic basis of functional health and frailty. *J. Gerontol. Biol. Sci.* **2002**, *57A*, B115–B125. [CrossRef]
17. Brink, K.J.; Kim, S.K.; Sommerfeld, J.H.; Amazeen, P.G.; Stergiou, N.; Likens, A.D. Pink noise promotes sooner state transitions during bimanual coordination. *Proc. Natl. Acad. Sci. USA* **2024**, *121*, e2400687121. [CrossRef]
18. Manor, B.; Lipsitz, L.A. Physiologic complexity and aging: Implications for physical function and rehabilitation. *Prog. Neuro-Psychopharmacol. Biol. Psychiatry* **2013**, *45*, 287–293. [CrossRef]
19. Dubois, D.M. Mathematical foundations of discrete and functional systems with strong and weak anticipations. *Lect. Notes Comput. Sci.* **2003**, *2684*, 110–132.
20. West, B.J.; Geneston, E.L.; Grigolini, P. Maximizing information exchange between complex networks. *Phys. Rep.* **2008**, *468*, 1–99. [CrossRef]
21. Stephen, D.G.; Stepp, N.; Dixon, J.A.; Turvey, M.T. Strong anticipation: Sensitivity to long-range correlations in synchronization behavior. *Phys. A* **2008**, *387*, 5271–5278. [CrossRef]
22. Thaut, M.H.; Abiru, M. Rhythmic auditory stimulation in rehabilitation of movement disorders: A review of current research. *Music. Percept.* **2010**, *27*, 263–269. [CrossRef]
23. Ghai, S.; Ghai, I.; Effenberg, A.O. Effect of rhythmic auditory cueing on aging gait: A systematic review and meta-analysis. *Aging Dis.* **2018**, *9*, 901. [CrossRef] [PubMed]
24. Hove, M.J.; Suzuki, K.; Uchitomi, H.; Orimo, S.; Miyake, Y. Interactive rhythmic auditory stimulation reinstates natural $1/f$ timing in gait of Parkinson's patients. *PLoS ONE* **2012**, *7*, e32600. [CrossRef] [PubMed]
25. Kaipust, J.P.; McGrath, D.; Mukherjee, M.; Stergiou, N. Gait variability is altered in older adults when listening to auditory stimuli with differing temporal structures. *Ann. Biomed. Eng.* **2013**, *41*, 1595–1603. [CrossRef]
26. Rhea, C.K.; Kiefer, A.W.; D'Andrea, S.E.; Warren, W.H.; Aaron, R.K. Entrainment to a real time fractal visual stimulus modulates fractal gait dynamics. *Hum. Mov. Sci.* **2014**, *36*, 20–34. [CrossRef]
27. Vaz, J.R.; Knarr, B.A.; Stergiou, N. Gait complexity is acutely restored in older adults when walking to a fractal-like visual stimulus. *Hum. Mov. Sci.* **2020**, *74*, 102677. [CrossRef]
28. Raffalt, P.C.; Sommerfeld, J.H.; Stergiou, N.; Likens, A.D. Stride-to-stride time intervals are independently affected by the temporal pattern and probability distribution of visual cues. *Neurosci. Lett.* **2023**, *792*, 136909. [CrossRef]

29. Jordao, S.; Cortes, N.; Gomes, J.; Brandao, R.; Santos, P.; Pezarat-Correia, P.; Oliveira, R.; Vaz, J. Synchronization performance affects gait variability measures during cued walking. *Gait Posture* **2022**, *96*, 351–356. [CrossRef]
30. Vaz, J.R.; Cortes, N.; Gomes, J.S.; Reis, J.F.; Stergiou, N. Stride-to-stride variability is altered when running to isochronous visual cueing but remains unaltered with fractal cueing. *Sports Biomech.* **2024**, 1–13. [CrossRef]
31. Marmelat, V.; Torre, K.; Beek, P.J.; Daffertshofer, A. Persistent fluctuations in stride intervals under fractal auditory stimulation. *PLoS ONE* **2014**, *9*, e91949. [CrossRef] [PubMed]
32. Roerdink, M.; Daffertshofer, A.; Marmelat, V.; Beek, P.J. How to sync to the beat of a persistent fractal metronome without falling off the treadmill? *PLoS ONE* **2015**, *10*, e0134148. [CrossRef]
33. Kiriella, J.B.; Di Bacco, V.E.; Hollands, K.L.; Gage, W.H. Evaluation of the effects of prescribing gait complexity using several fluctuating timing imperatives. *J. Mot. Behav.* **2020**, *52*, 570–577. [CrossRef] [PubMed]
34. Rhea, C.K.; Kiefer, A.W.; Wittstein, M.W.; Leonard, K.B.; MacPherson, R.P.; Wright, W.G.; Haran, F.J. Fractal gait patterns are retained after entrainment to a fractal stimulus. *PLoS ONE* **2014**, *9*, e106755. [CrossRef] [PubMed]
35. Uchitomi, H.; Ota, L.; Ogawa, K.-I.; Orimo, S.; Miyake, Y. Interactive rhythmic cue facilitates gait relearning in patients with Parkinson's disease. *PLoS ONE* **2013**, *8*, e72176. [CrossRef] [PubMed]
36. Damouras, S.; Chang, M.D.; Sejdic´, E.; Chau, T. An empirical examination of detrended fluctuation analysis for gait data. *Gait Posture* **2010**, *31*, 336–340. [CrossRef]
37. Manor, B.; Costa, M.D.; Hu, K.; Newton, E.; Starobinets, O.; Kang, H.G.; Peng, C.; Novak, V.; Lipsitz, L.A. Physiological complexity and system adaptability: Evidence from postural control dynamics of older adults. *J. Appl. Physiol.* **2010**, *109*, 1786–1791. [CrossRef]
38. Stergiou, N.; Decker, L.M. Human movement variability, nonlinear dynamics, and pathology: Is there a connection? *Hum. Mov. Sci.* **2011**, *30*, 869–888. [CrossRef]
39. Shahar, R.T.; Agmon, M. Gait analysis using accelerometry data from a single smartphone: Agreement and consistency between a smartphone application and gold-standard gait analysis system. *Sensors* **2021**, *21*, 7497. [CrossRef]
40. Hou, Y.; Wang, S.; Li, J.; Komal, S.; Li, K. Reliability and validity of a wearable inertial sensor system for gait assessment in healthy young adults. In Proceedings of the 14th International Congress on Image and Signal Processing, BioMedical Engineering and Informatics (CISP-BMEI), Shanghai, China, 23–25 October 2021; pp. 1–6.
41. Tack, G.R.; Choi, J.S. Validity of a portable APDM inertial sensor system for stride time and stride length during treadmill walking. *Korean J. Sport Biomech.* **2017**, *27*, 53–58. [CrossRef]
42. Washabaugh, E.P.; Kalyanaraman, T.; Adamczyk, P.G.; Claflin, E.S.; Krishnan, C. Validity and repeatability of inertial measurement units for measuring gait parameters. *Gait Posture* **2017**, *55*, 87–93. [CrossRef] [PubMed]
43. Gao, J.; Hu, J.; Tung, W.-W. Facilitating joint chaos and fractal analysis of biosignals through nonlinear adaptive filtering. *PLoS ONE* **2011**, *6*, e24331. [CrossRef] [PubMed]
44. Gao, J.; Sultan, H.; Hu, J.; Tung, W.-W. Denoising nonlinear time series by adaptive filtering and wavelet shrinkage: A comparison. *IEEE Signal Process. Lett.* **2010**, *17*, 237–240.
45. Hu, J.; Gao, J.; Wang, X. Multifractal analysis of sunspot time series: The effects of the 11-year cycle and Fourier truncation. *J. Stat. Mech. Theory Exp.* **2009**, *2009*, P02066. [CrossRef]
46. Gao, J.; Hu, J.; Mao, X.; Perc, M. Culturomics meets random fractal theory: Insights into long-range correlations of social and natural phenomena over the past two centuries. *J. R. Soc. Interface* **2012**, *9*, 1956–1964. [CrossRef]
47. Riley, M.A.; Bonnette, S.; Kuznetsov, N.; Wallot, S.; Gao, J. A tutorial introduction to adaptive fractal analysis. *Front. Physiol.* **2012**, *3*, 371. [CrossRef]
48. Lohse, K.R.; Lang, C.E.; Boyd, L.A. Is more better? Using metadata to explore dose–response relationships in stroke rehabilitation. *Stroke* **2014**, *45*, 2053–2058. [CrossRef]
49. Paula Quixadá, A.; de Castro, D.G.; Vivas Miranda, J.G. Scaling exponent of human gait: A scoping review. *Nonlinear Dyn. Psychol. Life Sci.* **2022**, *26*, 259–287.
50. Hollman, J.H.; Watkins, M.K.; Imhoff, A.C.; Braun, C.E.; Akervik, K.A.; Ness, D.K. A comparison of variability in spatiotemporal gait parameters between treadmill and overground walking conditions. *Gait Posture* **2016**, *43*, 204–209. [CrossRef]
51. Semaan, M.B.; Wallard, L.; Ruiz, V.; Gillet, C.; Leteneur, S.; Simoneau-Buessinger, E. Is treadmill walking biomechanically comparable to overground walking? A systematic review. *Gait Posture* **2022**, *92*, 249–257. [CrossRef]
52. White, S.C.; Yack, H.J.; Tucker, C.A.; Lin, H.-Y. Comparison of vertical ground reaction forces during overground and treadmill walking. *Med. Sci. Sports Exerc.* **1998**, *30*, 1537–1542. [CrossRef] [PubMed]

biomechanics

MDPI

Article

Reliability and Validity of the Articulation Motion Assessment System Using a Rotary Encoder

Hiroki Ito [1], Hideaki Yamaguchi [2], Mari Inoue [3], Hikaru Nagano [4], Ken Kitai [1], Kiichiro Morita [5] and Takayuki Kodama [1,5,*]

1 Graduate School of Health Science, Kyoto Tachibana University, Yamashina-ku, Kyoto 607-8175, Japan; h901524007@st.tachibana-u.ac.jp (H.I.); h901523004@st.tachibana-u.ac.jp (K.K.)
2 CARETECH Plus, Nagoya 462-0847, Japan; caretech.plus.hy@gmail.com
3 Graduate School of Human Development and Environment, Kobe University, Kobe 657-8501, Japan; inouema@kobe-u.ac.jp
4 Faculty of Fiber Science and Engineering, Kyoto Institute of Technology University, Matsugasaki, Sakyo-ku, Kyoto 606-8585, Japan; nagano@kit.ac.jp
5 Cognitive and Molecular Research Institute of Brain Diseases, Kurume University, Kurume, Fukuoka 830-0011, Japan; kiichiro@kurume-u.ac.jp
* Correspondence: kodama-t@tachibana-u.ac.jp; Tel.: +81-075-574-4312

Abstract: This study aimed to validate the effectiveness of the Articulation Motion Assessment System (AMAS), a joint kinematic evaluation system, for clinical applications. AMAS enables synchronised measurement using neurophysiological indicators, overcoming laboratory setting limitations. We compared AMAS-based ankle joint kinematic evaluations, particularly the sagittal and frontal plane angles, with two-dimensional (2D) motion analysis to determine the validity and reliability of AMAS. Both AMAS and 2D motion analysis reliably detected significant differences in angles within the sagittal and frontal planes. Correlation analysis revealed a significant moderate-to-strong correlation between the AMAS and the conventional method of 2D motion analysis, proving the measurement validity of the AMAS ($\rho = 0.53$–0.77 for sagittal plane angles; $\rho = 0.46$–0.72 for frontal plane angles). The average root mean squared error (RMSE) was significantly lower in AMAS ($10.90 \pm 2.93°$ for sagittal plane angles; $13.44 \pm 1.09°$ for frontal plane angles) than in the inertial sensor-based three-dimensional (3D) motion analysis. Reliability analysis revealed high reliability of measurements (intraclass correlation coefficients (ICC) ≥ 0.76). However, the Bland–Altman analysis identified a slightly lower fixed bias, which was observed as a characteristic of each measurement system. The AMAS accurately detects ankle joint angles without being constrained by measurement environment limitations. Synchronised measurements using neurophysiological indicators potentially contribute to understanding ankle joint control mechanisms and developing rehabilitation strategies.

Keywords: ankle motion; rotary encoder; multiple system; novel system; motion analysis

1. Introduction

Human walking is a uniquely acquired upright bipedal gait used to complete daily activities [1]. Walking maintains functional independence [2] and quality of life [3]. Typical clinical evaluation indexes of gait performance include gait speed, which is affected by gait pattern [4,5]. The control of the ankle joint is essential in regulating changes in walking speed by altering the gait pattern; however, age-related neurodegeneration interferes with this motor control, causing motor impairments and other neurological symptoms [6–8].

Therefore, assessing ankle joint control is crucial for rehabilitation, including attaining or restoring walking ability. The ankle joint control coordinates the relative timing of ankle joint torque exertion and angular displacement during changes in walking velocity [9]. It encompasses the multifaceted elements of the so-called movement in kinematics and its underlying neural mechanisms in neurophysiology. This results in numerous degrees of freedom due to anatomical, biomechanical, and neurophysiological redundancies within the interaction between the body and environment [10,11]. However, evaluating ankle joint control has been limited to one aspect of biomechanics [12,13] and neurophysiology [14] using electroencephalography and electromyography. Therefore, analysing the motor control of the ankle joint should include correlating neural activity with motor behaviour to understand the precise objectives of the central nervous system and the mechanisms by which it modulates the final output of motor actions [15,16]. Therefore, motor control should be comprehensively understood based on a unified theoretical framework [17].

Muscle synergies [18] involve the cooperative activities of multiple muscles and degrading and controlling movement. Additionally, neuromuscular control of the lower extremity by the upper motor centre (hereafter referred to as 'lower-extremity muscle control') [19] interacts with the body and the environment and influences muscle synergies. Understanding that lower-extremity muscle is being controlled by a nervous system-based motor control mechanism [20,21] will help clarify the neural mechanisms of movement disorders to capture the control mechanisms of various motor components [22–24]. Cortical–muscular coherence, a neurophysiological index for capturing neural mechanisms in lower-extremity muscle control, correlates brain activity with muscle activity [19]. It is a biomarker for movement disorders used in rehabilitation therapy [25]. However, in clinical application, cortical–muscular coherence on the affected side must be classified accurately by determining its relationship to movements hampered by ageing or disability to better understand its impact on movement disorders [26]. Therefore, neurorehabilitation for ankle joint control warrants elucidating human walking ability using the neurophysiological indices of lower-extremity muscle control. Due to its anatomical structure, ankle motion is a simultaneous and complex multi-axis movement [27], with significant individual differences and some even result in altered motion on the forefoot plane [28,29]. Therefore, we hypothesised that a new lower-extremity muscle control evaluation method combining kinematic and neurophysiological indices may capture ankle joint control, which is essential for gait.

This hypothesis requires measurements of synchronous joint motion, muscular activity, and brain activity. Therefore, a new system is necessary to measure ankle joint control kinematic and conventional neurophysiological indices and determine the relationship between each index. Electroencephalography (EEG) and electromyography (EMG) use non-invasive scalp or skin electrodes for neurophysiological analysis, which can be monitored remotely [30]. A 3D motion analysis using optical motion capture with markers is the gold standard for the kinematic analysis of walking and lower limb motion [31], despite the 3D motion analysis being expensive, requiring dedicated space [32], being time-consuming, and highly dependent on the evaluator's skill and experience. Since it involves manually identifying specific anatomical indicators captured by multiple videos [33], the measurement technique, environment, specifications, and number of cameras affect the accuracy of joint angle measurements [34]. A 2D motion analysis, which shows moderate correlation and high concordance with a 3D motion analysis [35], can help identify kinematics [36] and is used in foot kinematic analysis [37]. However, capturing optical motions using these markers requires video cameras and a restricted measurement environment.

Inertial sensor-based 3D motion analysers offer a simple method for outdoor measurement without a video camera. However, they have measurement errors due to magnetic field effects [38,39]. A measurement method without magnetic sensors has been proposed; however, the magnetic sensor information is required during initial calibration [40] with an environment free from magnetic field effects. Therefore, establishing an evaluation method integrating kinematic and neurophysiological indices remains challenging in terms of convenience and versatility in clinical applications.

We developed a rotary encoder-based ankle joint angle measurement system and an articulation motion assessment system (AMAS) to address the aforementioned clinical problems. This rotary encoder converts the mechanical displacement of rotation into an electrical signal and processes it to detect position, speed, and angle. It measures motor displacement in patients with Parkinson's [41] and in wearable exoskeleton robots supporting human movement [42]. Without video cameras or magnetic sensors, this system analyses the kinematics of dorsiflexion, plantar flexion, inversion, and eversion of the ankle. Additionally, AMAS synchronises with existing bio-signal processing devices that capture neurophysiological indicators, such as EEG and EMG. Synchronised ankle kinematic and neurophysiological analysis facilitates exploring the indices of these functional couplings for motor control. This enables elucidating mechanisms of ankle joint control, which is essential for walking.

This study aimed to establish the ankle motion task as a simple method of motion analysis using the newly developed AMAS and confirmed the validity and reliability of this system via a comparative validation with existing evaluations.

2. Materials and Methods

2.1. Participants

This study included 14 participants (12 men and two women; age 22 $\pm$ 1 years; height 1.70 $\pm$ 0.06 m; weight 65.07 $\pm$ 14.81 kg) enrolled between December 2022 and February 2023. The inclusion criteria were an absence of cognitive problems, motor sensory dysfunction in the lower extremities, and visual impairment. Fourteen healthy adults (twelve males and two females) participated in and completed the entire study. This study was explained verbally and in writing, and the participants provided written informed consent before participating. This study was approved by the Research Ethics Committee of Kyoto Tachibana University (Approval No. 22-13) and conducted in accordance with the Declaration of Helsinki.

2.2. Protocol

Each measurement device was attached to the participant before any measurements were collected (Figure 1a). This study used AMAS, Inertial Measurement Unit (IMU), and a 2D motion analysis to measure voluntary movements in the sagittal and frontal plane of the ankle joint. Voluntary movements often face redundancy issues [11], making it challenging to capture motor control due to the many degrees of freedom involved, especially in free and coarse movements [43,44]. Analytical methods have been employed to minimise these degrees of freedom and constrain the analysis to a single plane in two dimensions [45–47]. A 2D motion analysis is a widely used technique in clinical practice as an alternative to a 3D motion analysis, a costly and environmentally restricted technique [35–37]. It is also suitable for validation because it increases the measurement's simplicity and reproducibility by limiting it to a single plane [48]. The present study was designed to compare the developed AMAS with the IMU, which has been used in recent years, using a 2D motion analysis as a reference and an alternative method in biomechanics and rehabilitation.

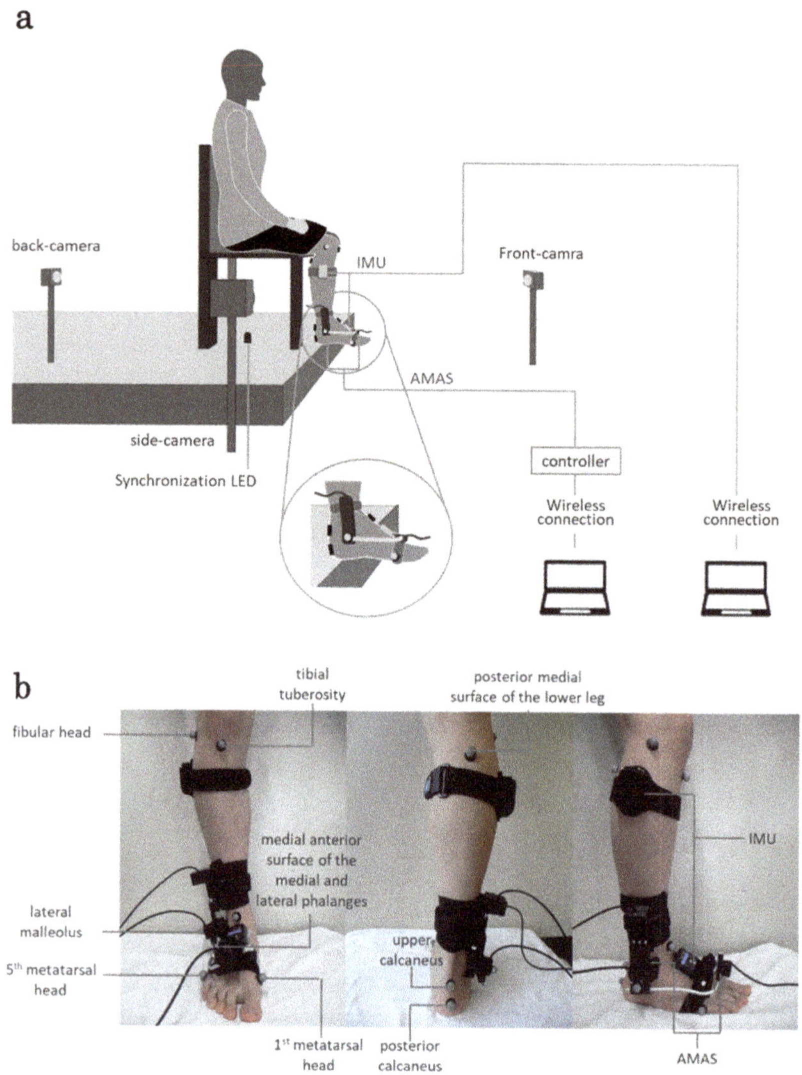

Figure 1. Experimental view. Video cameras were set up to capture the participant's right foot from the side, front, and rear, along with the synchro light (**a**). An actual magnified image is shown in (**b**). The AMAS and IMU are attached to the right lower extremity with a belt, and nine reflective markers are attached to the designated anatomical landmarks using double-sided tape.

All the tasks were performed with the participant sitting on a chair with a backrest one step above the floor. The measurement tasks consisted of automatic voluntary movements of dorsiflexion, plantar flexion, inversion, and eversion of the ankle on one side. They were performed in two ranges of motion, the maximum range of motion and the mild range of motion, according to the participant's subjective judgement. For each automatic voluntary movement, the following conditions were assigned according to the participant's subjective movement speed: 'movement at a comfortable speed' and 'movement at a slow speed'. These conditions were set to verify whether it was possible to measure a wide range of participants with mild to severe disabilities since the range of joint motion was assumed to be smaller and patients with motor dysfunction exhibited slow movement. Each participant was required to complete five consecutive measurement tasks. The measurement task sequence was randomised using a random number table. Each measuring device was calibrated simultaneously in the midfoot position. Only the heel is in contact with the floor, while the area distal to the heel is raised off the ground to maintain freedom of ankle motion. Because biomechanical and neurophysiological studies have shown that ankle motion was not lateralised [49,50], the kicking foot was determined to be the dominant foot and was used as the measurement limb. All the participants' dominant feet were determined to be the right foot. Measurements were taken using a video camera. The centre of the video camera lens was aligned to the centre of the foot; therefore, the height was adjusted using a tripod, and the camera was positioned 1 cm from the front and rear and 1 m from the right side.

The synchro light at the start of the IMU measurement and sensor mat light input to the AMAS were used to synchronise the data. The data's starting point was then temporally synchronised by the sensor mat signal input after initiating the AMAS measurement, and the video camera captured the sensor mat light. Synchronously measured data were corrected to align the data points, where time-series data of joint angles were recorded over five consecutive trials. All data acquired were used in this analysis, corrected for the 50 Hz sampling rate.

2.3. Instrumentation

An AMAS, reflective marker, and IMU sensor were attached to one of the participant's legs (Figure 1b).

2.3.1. AMAS

This joint angle measurement device simultaneously measures the plantarflexion/dorsiflexion and inversion/eversion angles during ankle joint motion and consists of an ankle joint orthosis, controller, and operation application (Figure 2a).

The ankle joint unit (Figure 2b) comprises two units with a built-in rotary encoder (manufactured by Supertech Electronic, Rixin Street, Chiayi City, Taiwan). One unit was attached to the outer lower leg, which served as the fundamental axis, and the other unit was attached to the dorsal foot (metatarsal head), which served as the moving axis.

A pipe connects both units, and plantar dorsiflexion movement is transmitted to the basic axis-side unit via the connecting pipe, causing the rotary encoder to rotate. Similarly, the basic shaft-side unit supported the connecting pipe, and the inversion/eversion motion of the ankle was transmitted to the rotary encoder of the mobile shaft-side unit. The controller transmits the signal from the rotary encoder, resulting in ankle motion, to the application (tablet). The application converts the signals into angles and displays the changes in plantarflexion/dorsiflexion and inversion/eversion angles in real time. The angle data were saved as a CSV file. The system simultaneously extracts the dorsiflexion/plantarflexion and inversion/eversion angles (degrees) of the ankle. The sampling rate was 50 Hz.

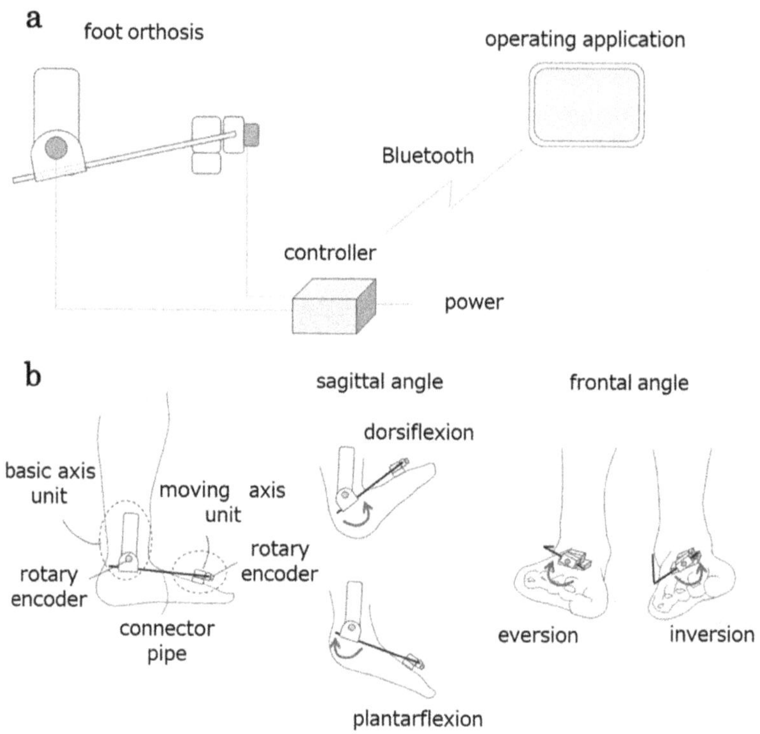

Figure 2. Configuration of the AMAS. The AMAS comprises three components: a foot orthosis, a controller wirelessly connected to the operating application, and foot orthosis (**a**). The foot orthosis measures the angles of the sagittal and frontal planes of the foot using a unit incorporating a rotary encoder on the basic axis and moving axis side (**b**).

2.3.2. A 2D Motion Analysis

The reference motion capture was performed using the free 2D motion analysis software Image J Version 1.54h (NIH, Bethesda, MD, USA) for a 2D motion analysis with reflective markers. The following nine anatomical landmarks were selected according to previous studies: fibular head; tibial tuberosity; posterior medial surface of the lower leg; medial anterior surface of the medial and lateral phalanges; lateral malleolus; first and fifth metatarsal heads; posterior calcaneus (CAL); and, directly above that, CAL [51]. Reflective markers were attached to the skin using double-sided tape at selected landmarks. Two 30 bits per second digital high-definition video cameras (Panasonic, Tokyo, Japan) were used to capture images from the anterior and posterior sides for the frontal plane task and the right lateral side for the sagittal plane task during the measurement task. The video captured by the video camera was clipped at 30 Hz using the free video playback software GOM Player 2.3.104.5374 (Gretech, Soul, Korea). Each data point was then extracted and corrected every 0.14 s. The captured data were used to calculate joint angles from 2D coordinates using the free 2D motion analysis software ImageJ. The dorsiflexion/plantarflexion angle of the ankle in the sagittal plane is the angle between the straight line connecting the fibular head and external capsule and the straight line connecting the fifth metatarsal head and external capsule. The inversion/eversion angle of the ankle viewed from the anterior is the angle between the lower leg axis consisting of the centre of the medial and lateral phalanges and the tibial coarse surface and the plantar axis consisting of the first and fifth metatarsal heads, based on the 'Methods for Indication and Measurement of Range of Motion of Joints'. The angle of inversion/eversion of the ankle viewed from the posterior is the angle between the line passing from the centre of the lower leg through the centre of the Achilles tendon and the line connecting the two markers at the rear of the heel.

The joint angles (degrees) were extracted from the 2D motion analysis of sagittal plane videos, with the dorsiflexion direction being positive and the plantar flexion direction being negative. The joint angles (degrees) were extracted from the 2D motion analysis of the frontal plane video, with the inversion direction being positive and the eversion direction being negative.

2.3.3. Inertial Measurement Unit (IMU)

Two MyoMotion sensors (Noraxon USA Inc., Scottsdale, AZ, USA) were used as IMU. The tibial sensor was wrapped around the lower leg using an attached belt, and the dorsal foot sensor was attached to double-sided tape. The sampling rate was 100 Hz.

From the IMU, dorsiflexion/plantarflexion and the ankle's internal/external return angles (degrees) were extracted using sensors placed across the ankle joint.

2.4. Data Analysis

The statistical software SPSS (version 27.0; IBM Corporation, Armonk, NY, USA) was used for statistical analysis, where the *p*-value was set at 5%. Corresponding *t*-tests were used to determine the time (s) difference between comfortable and slow conditions in the measurement task. All data were subjected to the Kolmogorov–Smirnov test of normality. The feature extraction of the AMAS, the regression model's performance, and the measurement method's reliability and agreement were statistically validated [52]. For assessing validity, comparisons between conditions were performed using the mean angles measured by each system and analysed with paired *t*-tests. In addition, the correlation coefficients between the AMAS and IMU were calculated from the time-series data of joint angles of the 2D motion analysis. Root mean squared error (RMSE) (1) was calculated as the performance index of the regression model using Equation (1) [53]:

$$\mathrm{RMSE} = \sqrt{\frac{\Sigma(\mathrm{X(t)} - \mathrm{Y(t)})^2}{\mathrm{n}}} \tag{1}$$

X is the joint angle measured by the 2D motion analysis as the reference value, and Y is calculated by substituting the measured joint angle by AMAS or IMU, squaring the difference for n data and taking the square root.

The maximum value of each of the five repeatedly extracted angular data was used to evaluate reliability. A two-way mixed model by a single ICC examiner was used [54]. Bland–Altman analysis was performed to confirm the presence of systematic errors in the overall data to evaluate the agreement between the measurement methods and the characteristics of the measurement methods using the margin of error (LOA) (2) [55].

$$\mathrm{LOA} = \frac{1}{n}\sum_{i=1}^{n} di \pm 1.96 \times \sqrt{\frac{1}{n-1}\sum_{i=1}^{n}\left(di - \overline{d}\right)^2} \tag{2}$$

3. Results

The time taken for each condition in the measurement task was 9.22 ± 2.99 s (AVE ± SD) in the comfortable condition and 13.64 ± 6.60 s (AVE ± SD) in the slow condition, with a significant difference ($t = -8.28$, $p < 0.001$) and 95% confidence interval

(CI): ([−5.49) to [−3.35]). It was confirmed that the joint angle data per measurement task extracted by each measuring device were non-normal using a normality test ($p < 0.01$).

3.1. Validity

3.1.1. Comparisons Between Conditions

In the sagittal plane angles, the comparison between the maximum and mild conditions revealed significant differences across all systems: a 2D motion analysis (t(13) = 9.523, $p < 0.001$); AMAS (t(13) = 10.498, $p < 0.001$); and IMU (t(13) = 2.708, $p = 0.018$). In contrast, for subjective motor speed, significant differences were found in a 2D motion analysis (t(13) = 4.142, $p = 0.001$) and AMAS (t(13) = 5.287, $p < 0.001$), but not in IMU (t(13) = 1.758, $p = 0.102$).

In the frontal plane angles, the comparison between the maximum and mild conditions showed significant differences in the 2D motion analysis performed from the front (t(13) = 6.597, $p < 0.001$), the 2D motion analysis performed from the back (t(13) = 4.661, $p < 0.001$), and AMAS (t(13) = 6.166, $p < 0.001$). However, no significant differences were observed in IMU (t(13) = 2.076, $p = 0.058$). In contrast, for subjective motor speed, significant differences were observed only in the 2D motion analysis performed from the back (t(13) = 2.816, $p = 0.015$). No significant differences were observed in the 2D motion analysis performed from the front (t(13) = 0.189, $p = 0.853$), AMAS (t(13) = −1.139, $p = 0.275$), or IMU (t(13) = 1.440, $p = 0.173$).

3.1.2. Criterion Validity

Spearman's correlation analysis was used to compute the correlation coefficients between the 2D motion analysis, AMAS, and IMU. In the sagittal plane angle, the correlation coefficient for the AMAS was ρ = 0.45 to 0.72 (Table 1). In the frontal plane angle, the correlation coefficient with the 2D motion analysis performed from the front was ρ = 0.39 to 0.72 (Table 2), and that with the 2D motion analysis performed from behind was ρ = −0.05 to 0.52 (Table 3). The correlation coefficient with the 2D motion analyses performed from the sagittal and front indicated a more than moderate correlation, while those with the 2D motion analysis from behind indicated a low correlation.

Table 1. Spearman's correlation coefficients and RMSE values between systems for sagittal.

					Subjective Motor Speed			
					Comfortable		**Slow**	
Activity	**Conditions**	**System**	**ρ**	**RMSE**	**ρ**	**RMSE**	**ρ**	**RMSE**
Dorsiflexion	maximum	AMAS	0.69 **	9.22	0.65 **	9.47	0.72 **	9.04
		IMU	0.44 **	17.3	0.41 **	22.23	0.47 **	12.73
	mild	AMAS	0.54 **	7.29	0.50 **	7.35	0.58 **	7.25
		IMU	0.35 **	10.19	0.28 **	10.02	0.40 **	10.31
Plantarflexion	maximum	AMAS	0.53 **	16.23	0.66 **	14.32	0.45 **	17.33
		IMU	0.47 **	21.26	0.38 **	22.34	0.54 **	20.55
	mild	AMAS	0.53 **	9.95	0.59 **	9.97	0.48 **	9.94
		IMU	0.16 **	19.66	0.17 **	17.12	0.16 **	21.23

** $p < 0.01$.

Table 2. Spearman's correlation coefficients and RMSE values between systems for 2D motion analysis viewed from the front.

					Subjective Motor Speed			
					Comfortable		Slow	
Activity	Conditions	System	ρ	RMSE	ρ	RMSE	ρ	RMSE
Inversion	maximum	AMAS	0.69 **	14.49	0.72 **	14.64	0.68 **	14.38
		IMU	0.02	28.82	−0.08 *	27.94	0.08 **	29.42
	mild	AMAS	0.55 **	13.11	0.63 **	12.29	0.49 **	13.63
		IMU	0.03	28.43	−0.07 *	29.78	0.10 **	27.49
Eversion	maximum	AMAS	0.46 **	14.74	0.39 **	15.29	0.52 **	14.31
		IMU	0.01	35.7	0.01	38.23	0	33.72
	mild	AMAS	0.50 **	11.19	0.55 **	11.15	0.46 **	11.22
		IMU	−0.06 *	34.96	−0.02	41.86	−0.07 *	28.77

* $p < 0.05$, ** $p < 0.01$.

Table 3. Spearman's correlation coefficients and RMSE values between systems for 2D motion analysis viewed from behind.

					Subjective Motor Speed			
					Comfortable		Slow	
Activity	Conditions	System	ρ	RMSE	ρ	RMSE	ρ	RMSE
Inversion	maximum	AMAS	0.48 **	17.46	0.50 **	17.8	0.48 **	17.22
		IMU	0.18 **	24.69	0.19 **	22.44	0.18 **	26.16
	mild	AMAS	0.49 **	12.89	0.52 **	12.75	0.49 **	12.98
		IMU	0.31 **	26.32	0.23 **	27	0.35 **	25.85
Eversion	maximum	AMAS	0.07 **	12.02	0.04	12.5	0.11 **	11.65
		IMU	0.02	27.49	0.13 **	28.7	−0.09 **	26.55
	mild	AMAS	−0.05 *	11.83	−0.07 *	12.49	−0.04	11.32
		IMU	0.04	28.17	0.11 **	34.04	−0.03	22.84

* $p < 0.05$, ** $p < 0.01$.

In the sagittal plane angle, the correlation coefficient for the IMU was ρ = 0.16 to 0.54 ($p < 0.01$). In the frontal plane angle, the correlation coefficient with the 2D motion analysis performed from the front was ρ = −0.08 to 0.10, and that with the 2D motion analysis performed from behind was ρ = −0.09 to 0.35 (Table 3). The correlation coefficient with the 2D motion analysis performed from the sagittal indicated a more than low correlation, while those with the 2D motion analyses from the front and behind indicated low correlations.

3.1.3. Root Mean Squared Error (RMSE)

The mean and standard deviation of RMSEs with each measurement device (AMAS and IMU) for the 2D motion analysis were calculated. The RMSE of the AMAS in the sagittal and frontal plane angle was lower than that of the IMU (Table 1, Table 2, Table 3 and Table S1). The RMSEs of the AMAS and IMU significantly differed in the *t*-test ($p < 0.01$).

3.2. Reliability

The ICC [56] was used to evaluate reliability. Five repetitions were treated as a fixed effect, and the ICC (3,k) measured how consistently the same evaluator could measure the participant's ankle movement angle under the same conditions. The angle measured per ankle movement task by the AMAS was above the ICC (3,1) of 0.76 under all conditions (Table 4).

Table 4. Reliability results ICC (3,1).

		Comfortable				Slow			
		ICC		95% CI		ICC		95% CI	
				Lower	Upper			Lower	Upper
Dorsiflexion	maximum	0.83	**	0.68	0.93	0.98	**	0.95	0.99
	mild	0.95	**	0.89	0.98	0.93	**	0.87	0.98
Plantarflexion	maximum	0.93	**	0.85	0.97	0.87	**	0.75	0.95
	mild	0.86	**	0.73	0.94	0.76	**	0.58	0.9
Inversion	maximum	0.96	**	0.91	0.98	0.98	**	0.96	0.99
	mild	0.97	**	0.94	0.99	0.97	**	0.94	0.99
Eversion	maximum	0.76	**	0.58	0.9	0.86	**	0.73	0.95
	mild	0.9	**	0.8	0.96	0.94	**	0.88	0.98

** $p < 0.01$.

Furthermore, ICC(3,5) was more significant than 0.91 under all conditions (Table 5).

Table 5. Reliability results ICC (3,5).

		Comfortable				Slow			
		ICC		95% CI		ICC		95% CI	
				Lower	Upper			Lower	Upper
Dorsiflexion	maximum	0.96	**	0.91	0.99	1	**	0.99	1
	mild	0.99	**	0.98	1	0.99	**	0.97	0.99
Plantarflexion	maximum	0.98	**	0.97	0.99	0.97	**	0.94	0.99
	mild	0.97	**	0.93	0.99	0.94	**	0.87	0.98
Inversion	maximum	0.99	**	0.98	1	1	**	0.99	1
	mild	0.99	**	0.99	1	0.99	**	0.99	1
Eversion	maximum	0.94	**	0.87	0.98	0.97	**	0.93	0.99
	mild	0.98	**	0.95	0.99	0.99	**	0.97	1

** $p < 0.01$.

3.3. Consistency

Bland–Altman plots were constructed between the AMAS and IMU against 2D motion analysis performed from the front with high values for the lateral and correlation coefficients in the ankle dorsiflexion/plantarflexion and inversion/eversion tasks. The vertical axis (DIFF) is the difference between the 2D motion analysis and AMAS or IMU, and the horizontal axis (MMEAN) is the mean between the 2D motion analysis and AMAS or IMU. LOA represents the range within which the difference between two measurements is acceptable as an error. Bland–Altman plots of sagittal plane angles in the dorsiflexion/plantarflexion task did not converge within the LOA for both instruments (Figure 3a).

The AMAS had a mean value of 1.26, with an LOA of 23.58 to −21.07. However, the IMU had a mean of −1.82 and an LOA of 30.00 to −36.54. The Bland–Altman plots of the frontal plane angle in the inversion/eversion task did not converge within the LOA for either instrument (Figure 3b).

The mean difference between the two systems for the AMAS was 0.18, and the LOA was from 26.70 to −26.33. However, the IMU had a mean of −9.54 for the difference between the two systems, and the LOA was from 50.48 to −69.56. *T*-tests were used to confirm the presence of fixed errors. For each task, the AMAS and IMU for the 2D motion analysis differed significantly, with a two-tailed test and a probability of significance of <1% and 95% CI: (1.02–1.49) for the AMAS and ([−2.19] to [−1.46]) for IMU for the sagittal plane angle and ([−0.09] to [0.46]) for the AMAS and ([−10.16] to [−8.91]) for IMU for the frontal plane angle. A fixed error was suggested by the absence of 0 in the 95% CI, except for the angle of the frontal plane by the AMAS. Regression analysis was performed to confirm the presence of proportional errors. The probability of significance of the regression equation, with DIFF as the dependent variable and MMEAN as the independent variable, was <1% for all tasks for both measurement methods, indicating no proportional error.

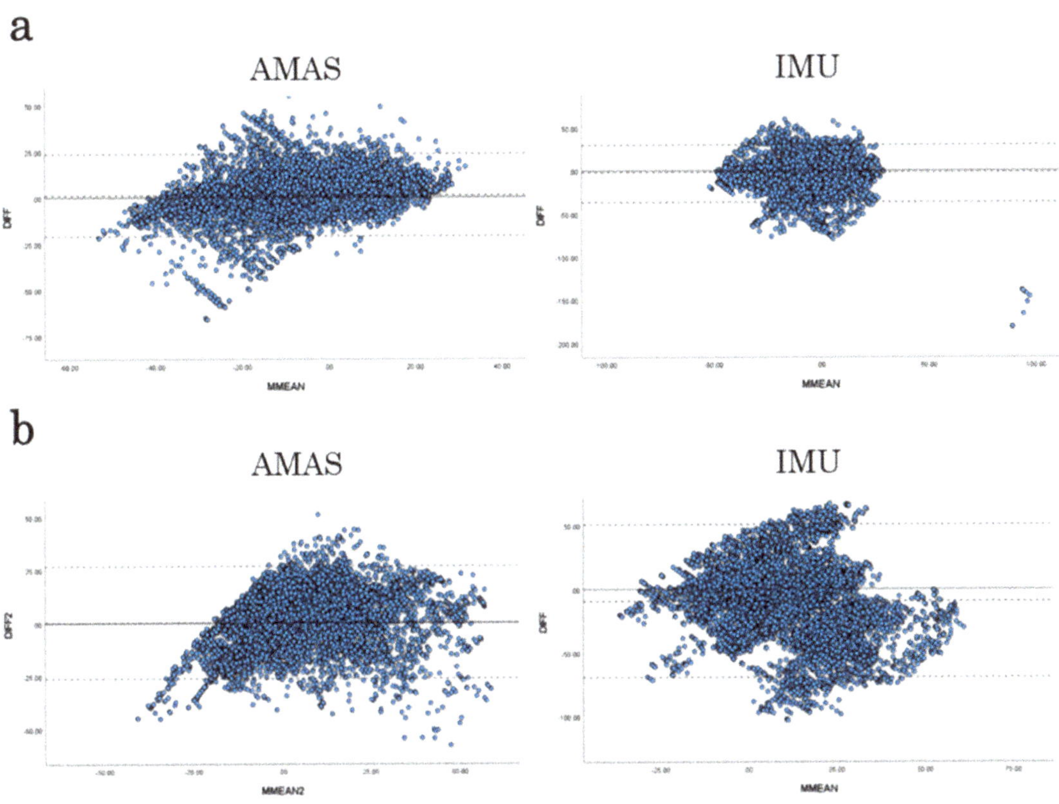

Figure 3. Bland–Altman plot. (**a**) The Bland–Altman plot showing the dorsiflexion and plantarflexion angles measured using the AMAS and IMU under all conditions. The solid line indicates 0, and the upper and lower dashed lines indicate LOA. (**b**) The Bland–Altman plot showing the ankle inversion and eversion angles measured by the AMAS and IMU under all conditions.

4. Discussion

We compared the reliability and validity of the AMAS in kinematic analysis with existing kinematic analysis assessment systems.

Comparisons between conditions demonstrated that the 2D Motion Analysis (both front and back) and AMAS consistently detected significant differences between most conditions. This indicates that both systems are generally reliable for evaluating sagittal and frontal plane angles. In particular, the 2D Motion Analysis and AMAS proved effective in detecting differences during maximum vs. mild conditions for both sagittal and frontal plane angles. In contrast, the IMU showed limitations in detecting differences, particularly in subjective motor speed and frontal plane angles. These limitations may stem from higher variability, lower resolution, or differences in data processing methods. As a result, the IMU appears less suited for identifying smaller differences or subtle changes, especially in subjective motor speed and frontal plane angle conditions. Criterion validation showed a significantly greater than moderate correlation between the AMAS and 2D motion analysis from the sagittal and anterior planes in all motor tasks and conditions without affecting joint angles and movement speed. The correlation coefficient values of 0–0.3, 0.3–0.7, and 0.7–1.0 represent a weak, moderate, and strong correlation, respectively [57]. Therefore, the present results revealed a more than moderate correlation. However, the IMU had no or a significantly low correlation with the 2D motion analysis. This suggests that the AMAS is more valid than the IMU for measuring the forefoot's sagittal and forehead angles. Outliers and fixation errors may have caused the lower validity of the IMU over the AMAS, which may be due to the magnetic effects. Biases exist in measurement systems, such as IMU, with an improved agreement with marker-based optical motion capture in the sagittal plane. Still, there are more significant errors in non-sagittal planes [58], and the differences vary among participants [59]. Standard wearable sensors' position, calibration method, and measurement algorithm affect accuracy [60]. Electromagnetic interference makes IMUs unreliable indoors [61]. Therefore, measurements should be collected in a magnetically neutral environment [62]. This study was conducted in a laboratory, and the results are consistent with previous studies. Our findings show that AMAS is a highly versatile system for clinical use.

Regarding the regression model performance index validation results, the mean RMSE of the AMAS based on the 2D motion analysis was significantly lower than that of the IMU. However, the mean RMSE of the AMAS was $10.90° \pm 2.93°$ for the sagittal angle and $13.44 \pm 1.09°$ for the frontal angle. Each participant's RMSE was 4.37–15.88° for the sagittal angle and 13.07–20.97° for the frontal angle, suggesting that significant differences in measurement angles were observed and errors were introduced. IMU kinematics should be interpreted cautiously [58,59] as the IMU must improve consistency and accuracy to replace marker-based optical 3D motion analysers, considered the gold standard. Our results confirm the limitations of joint kinematics analysis using IMUs owing to the magnetic effects.

Regarding the results of the reliability validation, the AMAS' reliability of sagittal and forehead angles had an ICC value of $\geq$0.76 for all motor tasks and conditions. This suggests that the reliability of measurements by the AMAS is higher than that by the conventional method [56]. Because obtaining kinematic data from individuals using the 3D motion analysis with markers, the gold standard method, is costly and complex, there is growing interest in a simple, inexpensive system with no local limitations [63]. The joint angles of ankle motion measured by this system are comparable to those obtained using previous measurement methods. This system has been verified to be a highly convenient measurement device because it solves the problem of environmental restrictions. The system's rotary encoder is a reliable and accurate sensor for measuring limb angles [64,65]; however, it is often attached to a rigid mechanical attachment on the exoskeleton and is, therefore, unsuitable as a wearable device [66]. Therefore, the external dimensions of this system were designed to be compact enough to be worn on the foot, increasing its versatility as a wearable device. However, regarding the results of the agreement validation, the Bland–Altman analysis showed that the differences between the 2D motion analysis and each of the measurement devices did not converge within the LOA, with no proportional error but a fixed error being found. The differences between the 2D motion analysis and AMAS were plotted upward in the sagittal and forehead planes, indicating that the 2D motion analysis tended to measure at lower values than those associated with the AMAS. However, the difference between 2D motion analysis and IMU was plotted downward in the sagittal and forehead planes, suggesting that IMU measurements tended to be higher than the 2D motion analysis measurements. The clinically acceptable joint angle error in gait motion analysis is <5° RMSE [67]. In the present study, the RMSE exceeded the clinically significant limit, although it differed from the gait motion. This raises the concern that while ankle joint angle measurement by AMAS can be used as a primary indicator, there may be some errors when comparing the ankle motion analysis data from other motion analysis devices.

A multidisciplinary team in collaboration with clinicians in the field is necessary to provide a diverse perspective [68]. The AMAS developed in our study measures ankle motion from two aspects, the sagittal plane and anterior frontal plane, and its high validity and high reliability have been verified using the measurement results from the 2D motion analysis. Additionally, the velocity of the ankle motion did not affect the system's angular measurement. Poor joint angle changes, decreased velocity in various movements, and abnormal compensatory movements accompany movement in post-stroke patients. Kinematic movement analysis results are essential for distinguishing between motor recovery and compensatory movement patterns [69,70]. Therefore, validation is needed in the rehabilitation field in patients presenting with various motor impairments, ranging from mild to severe.

In summary, AMAS is an evaluation system specialising in ankle joint motion control. Further validation can be applied to other joints and daily life activities. The development of AMAS is anticipated to contribute to the realisation of a multidimensional assessment of motor control in biomechanics and rehabilitation. It needs to be validated in other joints to demonstrate its usefulness as a wearable system.

The present study has several limitations. One limitation is that the results of this study measure non-weight-bearing motion that may not be valid or reliable during walking. Therefore, further analysis of actual walking motion measurements may be necessary before concluding that AMAS can measure motion under load, such as walking motion. Another limitation is that the subjects of this study were all healthy. The task of this study was to set up a condition of slow movement and minimum range of motion, assuming a patient, but we did not measure an actual patient. Therefore, further analysis of patients presenting with various movement disorders may be warranted in the future.

5. Conclusions

Because of convenience, versatility, and the ability to synchronously measure various biological signals, the AMAS was developed to assess ankle joint control, which is essential for walking in rehabilitation settings and real-life environments. The occurrence of a fixation error suggests room for improvement in this system. With future enhancements against errors, clinical evaluation using this system may help elucidate the mechanism of ankle joint control and develop individualised and optimised rehabilitation.

6. Practical Applications

The importance of capturing such quantitative analysis of human movement in free daily activities has been highlighted [71], and the AMAS has the potential to be applied to the evaluation of ankle joint motor control in various activities of daily living (ADLs). For example, it could assess movements in which ankle joint control is crucial, such as ascending and descending stairs, rising from a chair, and crossing an obstacle. However, it is necessary to optimise the measurement protocols and consider precautions for data interpretation according to the characteristics of each movement.

7. Patents

Japanese Patent Application No. 2023-67115.

Supplementary Materials: The following supporting information can be downloaded at https://www.mdpi.com/article/10.3390/biomechanics5010002/s1, Table S1. Spearman's correlation coefficient, RMSE values per participant. The correlation coefficient (ρ) and RMSE (°) for each participant from A to N are shown.

Author Contributions: H.I. and T.K. conceived and conceptualised this study. T.K. and H.Y. collected the data. H.I. designed and performed the statistical analysis. H.I. contributed to the interpretation of the results. H.I. prepared the original manuscript. T.K. oversaw the conduct of this study. H.Y., H.N., M.I., K.K., K.M., and T.K. were responsible for the development of this study. All authors reviewed this manuscript and revised it critically for intellectual content. All authors have read and agreed to the published version of the manuscript.

Funding: This research received no external funding.

Institutional Review Board Statement: This study was conducted in accordance with the Declaration of Helsinki and approved by the Institutional Review Board (or Ethics Committee) of Kyoto Tachibana University (Approval No. 22-13, 27 June 2022).

Informed Consent Statement: Written informed consent has been obtained from the patients to publish this paper.

Data Availability Statement: The authors will make the raw data supporting this article's conclusions available upon request.

Acknowledgments: I would like to thank T.K. for valuable discussions. I am grateful to H.Y., H.N., M.I., K.K., K.M., and T.K. for their assistance with the numerical simulations and for carefully proofreading this manuscript.

Conflicts of Interest: The authors declare no conflicts of interest.

References

1. Schmitt, D. Insights into the evolution of human bipedalism from experimental studies of humans and other primates. *J. Exp. Biol.* **2003**, *206*, 1437–1448. [CrossRef] [PubMed]
2. Mirelman, A.; Shema, S.; Maidan, I.; Hausdorff, J.M. Gait. *Handb. Clin. Neurol.* **2018**, *159*, 119–134. [CrossRef]
3. Baker, J.M. Gait disorders. *Am. J. Med.* **2018**, *131*, 602–607. [CrossRef]
4. Perry, J.; Garrett, M.; Gronley, J.K.; Mulroy, S.J. Classification of walking handicap in the stroke population. *Stroke* **1995**, *26*, 982–989. [CrossRef] [PubMed]
5. Schmid, A.; Duncan, P.W.; Studenski, S.; Lai, S.M.; Richards, L.; Perera, S.; Wu, S.S. Improvements in speed-based gait classifications are meaningful. *Stroke* **2007**, *38*, 2096–2100. [CrossRef] [PubMed]
6. Graf, A.; Judge, J.O.; Õunpuu, S.; Thelen, D.G. The effect of walking speed on lower-extremity joint powers among elderly adults who exhibit low physical performance. *Arch. Phys. Med. Rehabil.* **2005**, *86*, 2177–2183. [CrossRef] [PubMed]
7. Lamontagne, A.; Richards, C.L.; Malouin, F. Coactivation during gait as an adaptive behavior after stroke. *J. Electromyogr. Kinesiol.* **2000**, *10*, 407–415. [CrossRef] [PubMed]
8. Allen, J.L.; Kautz, S.A.; Neptune, R.R. The influence of merged muscle excitation modules on post-stroke hemiparetic walking performance. *Clin. Biomech.* **2013**, *28*, 697–704. [CrossRef] [PubMed]
9. Wade, L.; Birch, J.; Farris, D.J. Walking with increasing acceleration is achieved by tuning ankle torque onset timing and rate of torque development. *J. R. Soc. Interface* **2022**, *19*, 20220035. [CrossRef]
10. Morasso, P. A Vexing question in motor control: The degrees of freedom problem. *Front. Bioeng. Biotechnol.* **2022**, *9*, 783501. [CrossRef]

11. Bernstein, N.A. *The Coordination and Regulation of Movements*; Pergamon Press: London, UK, 1967; pp. 143–168.
12. Jenkyn, T.R.; Anas, K.; Nichol, A. Foot segment kinematics during normal walking using a multisegment model of the foot and ankle complex. *J. Biomech. Eng.* **2009**, *131*, 034504. [CrossRef]
13. Farris, D.J.; Kelly, L.A.; Cresswell, A.G.; Lichtwark, G.A. The functional importance of human foot muscles for bipedal locomotion. *Proc. Natl. Acad. Sci. USA* **2019**, *116*, 1645–1650. [CrossRef] [PubMed]
14. Yokoyama, H.; Kaneko, N.; Ogawa, T.; Kawashima, N.; Watanabe, K.; Nakazawa, K. Cortical correlates of locomotor muscle synergy activation in humans: An electroencephalographic decoding study. *iScience* **2019**, *15*, 623–639. [CrossRef]
15. Churchland, M.M.; Cunningham, J.P.; Kaufman, M.T.; Foster, J.D.; Nuyujukian, P.; Ryu, S.I.; Shenoy, K.V. Neural population dynamics during reaching. *Nature* **2012**, *487*, 51–56. [CrossRef] [PubMed]
16. Suresh, A.K.; Goodman, J.M.; Okorokova, E.V.; Kaufman, M.; Hatsopoulos, N.G.; Bensmaia, S.J.; Anatomy; Biology, Q.; Behavior, H.; States, U. Neural population dynamics in motor cortex are different for reach and grasp. *eLife* **2020**, *9*, 58848. [CrossRef] [PubMed]
17. McLoughlin, J. Ten guiding principles for movement training in neurorehabilitation. *OpenPhysio J.* **2020**, *10*, 1–17. [CrossRef]
18. Mehrabi, N.; Schwartz, M.H.; Steele, K.M. Can altered muscle synergies control unimpaired gait? *J. Biomech.* **2019**, *90*, 84–91. [CrossRef]
19. Petersen, T.H.; Willerslev-Olsen, M.; Conway, B.A.; Nielsen, J.B. The motor cortex drives the muscles during walking in human subjects. *J. Physiol.* **2012**, *590*, 2443–2452. [CrossRef]
20. Song, S.; Desai, R.; Geyer, H. Integration of an adaptive swing control into a neuromuscular human walking model. In Proceedings of the 2013 35th Annual International Conference of the IEEE Engineering in Medicine and Biology Society (EMBC), Osaka, Japan, 3–7 July 2013; pp. 4915–4918. [CrossRef]
21. Akazawa, K.; Fuj, K. Theory of muscle contraction and motor control. *Adv. Robot.* **1986**, *1*, 379–390. [CrossRef]
22. Takei, T.; Lomber, S.G.; Cook, D.J.; Scott, S.H. Transient deactivation of dorsal premotor cortex or parietal area 5 impairs feedback control of the limb in macaques. *Curr. Biol.* **2021**, *31*, 1476–1487.e5. [CrossRef]
23. Minassian, K.; Hofstoetter, U.S.; Dzeladini, F.; Guertin, P.A.; Ijspeert, A. The human central pattern generator for locomotion: Does it exist and contribute to walking? *Neuroscientist* **2017**, *23*, 649–663. [CrossRef]
24. Leiras, R.; Cregg, J.M.; Kiehn, O. Brainstem circuits for locomotion. *Annu. Rev. Neurosci.* **2022**, *45*, 63–85. [CrossRef]
25. Krauth, R.; Schwertner, J.; Vogt, S.; Lindquist, S.; Sailer, M.; Sickert, A.; Lamprecht, J.; Perdikis, S.; Corbet, T.; Millán, J.d.R.; et al. Cortico-muscular coherence is reduced acutely post-stroke and increases bilaterally during motor recovery: A pilot study. *Front. Neurol.* **2019**, *10*, 126. [CrossRef]
26. Liu, J.; Sheng, Y.; Liu, H. Corticomuscular coherence and its applications: A review. *Front. Hum. Neurosci.* **2019**, *13*, 100. [CrossRef]
27. Brockett, C.L.; Chapman, G.J. Biomechanics of the ankle. *Orthop. Trauma* **2016**, *30*, 232–238. [CrossRef]
28. Kim, H.; Cho, J.-E.; Seo, K.-J.; Lee, J. Bilateral ankle deformities affects gait kinematics in chronic stroke patients. *Front. Neurol.* **2023**, *14*, 1078064. [CrossRef] [PubMed]
29. Hong, W.-H.; Wang, C.-M.; Chen, C.-K.; Wu, K.P.-H.; Kang, C.-F.; Tang, S.F. Kinematic features of rear-foot motion using anterior and posterior ankle-foot orthoses in stroke patients with hemiplegic gait. *Arch. Phys. Med. Rehabil.* **2010**, *91*, 1862–1868. [CrossRef]
30. Palumbo, A.; Vizza, P.; Calabrese, B.; Ielpo, N. Biopotential signal monitoring systems in rehabilitation: A review. *Sensors* **2021**, *21*, 7172. [CrossRef]
31. Munro, A.; Herrington, L.; Carolan, M. Reliability of 2-dimensional video assessment of frontal-plane dynamic knee valgus during common athletic screening tasks. *J. Sport Rehabil.* **2012**, *21*, 7–11. [CrossRef]
32. Schurr, S.A.; Marshall, A.N.; Resch, J.E.; A Saliba, S. Two-dimensional video analysis is comparable to 3D motion capture in lower extremity movement assessment. *Int. J. Sports Phys. Ther.* **2017**, *12*, 163–172.
33. Corazza, S.; Mündermann, L.; Gambaretto, E.; Ferrigno, G.; Andriacchi, T.P. Markerless motion capture through visual hull, articulated ICP and subject specific model generation. *Int. J. Comput. Vis.* **2010**, *87*, 156–169. [CrossRef]
34. Simon, S.R. Quantification of human motion: Gait analysis—Benefits and limitations to its application to clinical problems. *J. Biomech.* **2004**, *37*, 1869–1880. [CrossRef]
35. Michelini, A.; Eshraghi, A.; Andrysek, J. Two-dimensional video gait analysis: A systematic review of reliability, validity, and best practice considerations. *Prosthet. Orthot. Int.* **2020**, *44*, 245–262. [CrossRef] [PubMed]
36. Dingenen, B.; Malliaras, P.; Janssen, T.; Ceyssens, L.; Vanelderen, R.; Barton, C.J. Two-dimensional video analysis can discriminate differences in running kinematics between recreational runners with and without running-related knee injury. *Phys. Ther. Sport* **2019**, *38*, 184–191. [CrossRef] [PubMed]
37. Cordova, M.L.; Dorrough, J.L.; Kious, K.; Ingersoll, C.D.; Merrick, M.A. Prophylactic ankle bracing reduces rearfoot motion during sudden inversion. *Scand. J. Med. Sci. Sports* **2007**, *17*, 216–222. [CrossRef] [PubMed]

38. Laidig, D.; Schauer, T.; Seel, T. Exploiting kinematic constraints to compensate magnetic disturbances when calcu-lating joint angles of approximate hinge joints from orientation estimates of inertial sensors. In Proceedings of the 2017 International Conference on Rehabilitation Robotics (ICORR), London, UK, 17–20 July 2017; pp. 971–976. [CrossRef]
39. McGrath, T.; Stirling, L. Body-Worn IMU-Based Human hip and knee kinematics estimation during treadmill walking. *Sensors* **2022**, *22*, 2544. [CrossRef]
40. Taetz, B.; Bleser, G.; Miezal, M. Towards self-calibrating inertial body motion capture. *Int. Conf. Inf. Fusion* **2016**, *19*, 1751–1759.
41. Chan, P.Y.; Ripin, Z.M.; Halim, S.A.; Tharakan, J.; Muzaimi, M.; Ng, K.S.; Kamarudin, M.I.; Eow, G.B.; Hor, J.Y.; Tan, K.; et al. An in-laboratory validity and reliability tested system for quantifying hand-arm tremor in motions. *IEEE Trans. Neural Syst. Rehabil. Eng.* **2018**, *26*, 460–467. [CrossRef]
42. Jaramillo, I.E.; Jeong, J.G.; Lopez, P.R.; Lee, C.-H.; Kang, D.-Y.; Ha, T.-J.; Oh, J.-H.; Jung, H.; Lee, J.H.; Lee, W.H.; et al. Real-time human activity recognition with IMU and encoder sensors in wearable exoskeleton robot via deep learning networks. *Sensors* **2022**, *22*, 9690. [CrossRef]
43. Yang, J.-F.; Scholz, J.P. Learning a throwing task is associated with differential changes in the use of motor abundance. *Exp. Brain Res.* **2005**, *163*, 137–158. [CrossRef] [PubMed]
44. Tseng, Y.-W.; Scholz, J.P.; Schöner, G.; Hotchkiss, L. Effect of accuracy constraint on joint coordination during pointing movements *Exp. Brain Res.* **2003**, *149*, 276–288. [CrossRef]
45. Gielen, C.; van Bolhuis, B.; Theeuwen, M. On the control of biologically and kinematically redundant manipulators. *Hum. Mov. Sci.* **1995**, *14*, 487–509. [CrossRef]
46. Koster, B.; Deuschl, G.; Lauk, M.; Timmer, J.; Guschlbauer, B.; Lücking, C.H. Essential tremor and cerebellar dysfunction: Abnormal ballistic movements. *J. Neurol. Neurosurg. Psychiatry* **2002**, *73*, 400–405. [CrossRef] [PubMed]
47. Matsugi, A.; Nishishita, S.; Bando, K.; Kikuchi, Y.; Tsujimoto, K.; Tanabe, Y.; Yoshida, N.; Tanaka, H.; Douchi, S.; Honda, T.; et al Excessive excitability of inhibitory cortical circuit and disturbance of ballistic targeting movement in degenerative cerebellar ataxia. *Sci. Rep.* **2023**, *13*, 13917. [CrossRef]
48. Mihcin, S. Simultaneous validation of wearable motion capture system for lower body applications: Over single plane range of motion (ROM) and gait activities. *Biomed. Eng. Biomed. Tech.* **2022**, *67*, 185–199. [CrossRef] [PubMed]
49. McCurdy, K.; Langford, G. Comparison of unilateral squat strength between the dominant and non-dominant leg in men and women. *J. Sports Sci. Med.* **2005**, *4*, 153–159. [PubMed]
50. Kapreli, E.; Athanasopoulos, S.; Papathanasiou, M.; Van Hecke, P.; Strimpakos, N.; Gouliamos, A.; Peeters, R.; Sunaert, S. Lateralization of brain activity during lower limb joints movement. An fMRI study. *NeuroImage* **2006**, *32*, 1709–1721. [CrossRef] [PubMed]
51. Mousavi, S.H.; Hijmans, J.M.; Moeini, F.; Rajabi, R.; Ferber, R.; van der Worp, H.; Zwerver, J. Validity and reliability of a smartphone motion analysis app for lower limb kinematics during treadmill running. *Phys. Ther. Sport* **2020**, *43*, 27–35. [CrossRef]
52. Choo, C.Z.Y.; Chow, J.Y.; Komar, J. Validation of the Perception Neuron system for full-body motion capture. *PLoS ONE* **2022**, *17*, e0262730. [CrossRef] [PubMed]
53. Mould, R.F. *Introductory Medical Statistics*, 3rd ed.; Boca Raton: London, UK, 1998; pp. 177–194.
54. Koo, T.K.; Li, M.Y. A guideline of selecting and reporting intraclass correlation coefficients for reliability research. *J. Chiropr. Med.* **2016**, *15*, 155–163. [CrossRef]
55. Bland, J.M.; Altman, D.G. Statistical methods for assessing agreement between two methods of clinical measurement. *Lancet* **1986**, *1*, 307–310. [CrossRef]
56. Brookshaw, M.; Sexton, A.; McGibbon, C.A. Reliability and validity of a novel wearable device for measuring elbow strength *Sensors* **2020**, *20*, 3412. [CrossRef]
57. Cohen, J. *Statistical Power Analysis for the Behavioral Sciences*, 2nd ed.; Routledge Academic, Lawrence Erlbaum Associates: New York, NY, USA, 1988; pp. 75–105.
58. Heuvelmans, P.; Benjaminse, A.; Bolt, R.; Baumeister, J.; Otten, E.; Gokeler, A. Concurrent validation of the Noraxon MyoMotion wearable inertial sensors in change-of-direction and jump-landing tasks. *Sports Biomech.* **2022**, *3*, 1–16. [CrossRef] [PubMed]
59. Rekant, J.; Rothenberger, S.; Chambers, A. Inertial measurement unit-based motion capture to replace camera-based systems for assessing gait in healthy young adults: Proceed with caution. *Meas. Sensors* **2022**, *23*, 100396. [CrossRef] [PubMed]
60. Walmsley, C.P.; Williams, S.A.; Grisbrook, T.; Elliott, C.; Imms, C.; Campbell, A. Measurement of upper limb range of motion using wearable sensors: A systematic review. *Sports Med. Open* **2018**, *4*, 53. [CrossRef] [PubMed]
61. McGrath, T.; Stirling, L. Body-Worn IMU Human skeletal pose estimation using a factor graph-based optimization framework *Sensors* **2020**, *20*, 6887. [CrossRef]
62. de Vries, W.; Veeger, H.; Baten, C.; van der Helm, F. Magnetic distortion in motion labs, implications for validating inertial magnetic sensors. *Gait Posture* **2009**, *29*, 535–541. [CrossRef]

63. Sprager, S.; Juric, M.B. Inertial sensor-based gait recognition: A review. *Sensors* **2015**, *15*, 22089–22127. [CrossRef]
64. Freeman, C.T.; Rogers, E.; Hughes, A.-M.; Burridge, J.H.; Meadmore, K.L. Iterative learning control in health care: Electrical stimulation and robotic-assisted upper-limb stroke rehabilitation. *IEEE Control. Syst.* **2012**, *32*, 18–43. [CrossRef]
65. Sharma, N.; Stegath, K.; Gregory, C.M.; Dixon, W.E. Nonlinear neuromuscular electrical stimulation tracking control of a human limb. *IEEE Trans. Neural Syst. Rehabil. Eng.* **2009**, *17*, 576–584. [CrossRef]
66. Allen, M.; Zhong, Q.; Kirsch, N.; Dani, A.; Clark, W.W.; Sharma, N. A nonlinear dynamics-based estimator for functional electrical stimulation: Preliminary results from lower-leg extension experiments. *IEEE Trans. Neural Syst. Rehabil. Eng.* **2017**, *25*, 2365–2374. [CrossRef]
67. McGinley, J.L.; Baker, R.; Wolfe, R.; Morris, M.E. The reliability of three-dimensional kinematic gait measurements: A systematic review. *Gait Posture* **2009**, *29*, 360–369. [CrossRef] [PubMed]
68. Kwakkel, G.; Van Wegen, E.; Burridge, J.; Winstein, C.; van Dokkum, L.; Murphy, M.A.; Levin, M.; Krakauer, J. Standardised measurement of quality of upper limb movement after stroke: Consensus-based core recommendations from the Second Stroke Recovery and Rehabilitation Roundtable. *Int. J. Stroke* **2019**, *14*, 783–791. [CrossRef]
69. Levin, M.F.; Kleim, J.A.; Wolf, S.L. What do motor "recovery" and "compensation" mean in patients following stroke? *Neurorehabilit. Neural Repair* **2008**, *23*, 313–319. [CrossRef]
70. Bernhardt, J.; Hayward, K.S.; Kwakkel, G.; Ward, N.S.; Wolf, S.L.; Borschmann, K.; Krakauer, J.W.; Boyd, L.A.; Carmichael, S.T.; Corbett, D.; et al. Agreed definitions and a shared vision for new standards in stroke recovery research: The Stroke Recovery and Rehabilitation Roundtable taskforce. *Int. J. Stroke* **2017**, *12*, 444–450. [CrossRef] [PubMed]
71. Maura, R.M.; Parra, S.R.; Stevens, R.E.; Weeks, D.L.; Wolbrecht, E.T.; Perry, J.C. Literature review of stroke assessment for upper-extremity physical function via EEG, EMG, kinematic, and kinetic measurements and their reliability. *J. Neuroeng. Rehabil.* **2023**, *20*, 21. [CrossRef]

biomechanics

MDPI

Article

A Comparison Between the Use of an Infrared Contact Mat and an IMU During Kinematic Analysis of Horizontal Jumps

Bjørn Johansen [1], Jono Neville [2] and Roland van den Tillaar [1,2,*]

[1] Department of Sports Sciences and Physical Education, Nord University, 7600 Levanger, Norway; bjorn.johansen2@student.nord.no
[2] Sports Performance Research Institute New Zealand, Auckland University of Technology, Auckland 1010, New Zealand; jono.neville@aut.ac.nz
* Correspondence: roland.v.tillaar@nord.no

Abstract: Background/Objectives: This study compared step-by-step kinematic measurements from an infrared contact mat (IR-mat) and an inertial measurement unit (IMU) system during bounding and single leg jumping for speed, while also evaluating the validity of algorithms originally developed for sprinting and running when applied to horizontal jumps. The aim was to investigate differences in contact times between the systems. Methods: Nineteen female football players (15 ± 0.5 years, 61.0 ± 5.9 kg, 1.70 ± 0.06 m) performed attempts in both jumps over 20 m with maximum speed, of which the first eight steps were analysed. Results: Significant differences were found between the systems, with the IR-mat recording longer contact times than the IMU. The IR-mat began and ended its measurements slightly earlier and later, respectively, compared to the IMU system, likely due to the IMU's algorithm, which was developed for sprinting with forefoot contact, while more midfoot and heel landing is used during jumps. Conclusions: Both systems provide reliable measurements; however, the IR mat consistently records slightly longer contact times for horizontal jumps. While the IMU is dependable, it exhibits a consistent bias compared to the IR mat. For bounding, the IR mat begins recording 0.018 s earlier at touch down and stops 0.021 s later. For single leg jumps, it starts 0.024 s earlier and ends 0.021 s later, resulting in contact times that are, on average, 0.039–0.045 s longer. These findings provide valuable insights for coaches and researchers in selecting appropriate measurement tools, highlighting the systematic differences between IR mats and IMUs in horizontal jump analysis.

Keywords: step kinematics; measurement accuracy; bounding; single leg jump

1. Introduction

In sports like horizontal jumping, spatiotemporal parameters are often used to identify different determinants for performance [1]. These parameters are important in various areas such as sports performance analysis, rehabilitation, and biomechanical research as they provide insights into both immediate performance metrics and long-term development trends [2,3]. Horizontal jumps, such as bounding and single leg jumps (plyometric exercises), are widely used to develop explosive strength in participants of various sports [4–6]. There are many technologies that enable the measurement of step kinematics, such as high-speed cameras [7], force plates [8], 3D motion capture systems [9,10], inertial measurement units (IMUs) [11–13], and infrared contact mats [14–16].

To measure multiple horizontal jumps with 3D systems and force plates, expensive equipment and time-consuming setup procedures are needed. Moreover, specialized expertise and access to high-performance laboratories are often required to conduct accurate analyses. In contrast, infrared mats or wearable inertial measurement unit (IMU) systems offer a feasible, user-friendly, portable, and cost-effective solution. An infrared optical contact mat creates a "carpet" of infrared light a few millimetres above the ground. When the light is interrupted, the unit records the timestamp of the event, allowing the calculation of both contact and flight times based on the differences between their timestamps. van den Tillaar [16] showed that the use of Optojump infrared contact mats for stride detection in sprinting was accurate compared to laser/IMU measurements, which is also supported by studies that have confirmed the high validity of OptoGait in spatiotemporal analyses [15,17]. Previous studies have validated the accuracy of the Optojump system against force plates and high speed cameras for sprint and jump kinematics [15,17]. However, in both these studies [15,17] the Optojump system was compared to a stationary system, such as a treadmill, or vertical jumps. Therefore, it is not known how accurate it would be for horizontal jumps. Moreover, van den Tillaar [14] found that the Musclelab infrared mat produces comparable kinematic measurements to Optojump during sprints. Based on this, it can be inferred that the Musclelab infrared mat provides measurements that are as accurate as those from force plates for sprinting. Similarly, the use of IMUs has been shown to reliably record stride patterns during walking, running, and sprinting [11,12,18,19]. In addition, van den Tillaar et al. [16] found that a laser gun combined with IMUs attached to each foot (Musclelab, Ergotest Technology AS, Langesund, Norway) can automatically detect accurate step-by-step kinematics comparable to those detected by force plates during high-speed sprinting. Therefore, it seems that contact mats and IMUs are systems that can accurately detect step kinematics during sprints. However, it has been established that differences in velocity and step characteristics are present between sprints and various horizontal jumps [20]. During sprint races, fast sprinters make first ground contact with the front part of the foot, reach over 10 m/s, and have contact times and flight times of around 0.1–0.12 s [4]. Bounding for speed has certain similarities to sprint running, including landing on the front foot, but involves somewhat lower speed and longer contact and flight times. Single leg jumps are relatively unlike sprint running, with the heel making first contact with the ground, speeds below 6 m/s, and contact and flight times closer to 0.2 and 0.4 s [4,21]. Due to these differences, the accuracy of new technologies with regards to the measurement of these kinematic variables must be investigated.

Therefore, the aim of this study was to compare measurements of step-by-step kinematics between infrared contact mat and an IMU during bounding and single leg jumps for speed. This study evaluates already-established algorithms, originally developed for sprinting and running, to investigate their validity and reliability when applied to horizontal jumps. Furthermore, we investigate whether measurements between the two systems are affected by different jumping tasks (bounding and single leg jumps).

2. Materials and Methods

2.1. Participants

Nineteen moderately experienced female football players (age 15 ± 0.5 years, body mass 61.0 ± 5.9 kg, body height 1.70 ± 0.06 m) from a local football club participated in this study. The participants, who were selected as part of a larger training study where this equipment was used to assess their performance, were familiar with the exercises, having practiced them for six weeks prior to the tests. The participants were fully informed about the procedures, the potential risks, and the benefits of the study, both in writing and orally. Written consent was obtained prior to all testing. Parental consent was obtained for all subjects The study was conducted following the latest revision of the Declaration of Helsinki and approved by the Norwegian Agency for Shared Services in Education and Research (project number 225529).

2.2. Procedure

Participants carried out an individual warm-up, consisting of approximately 5 min of jogging, followed by 5 min of an injury prevention program [22] and two submaximal sprints. After completing the warm-up, the participants performed a series of maximum-effort jump trials over a distance of 20 m. This distance was selected to ensure that all phases of the jump sequence—acceleration and maximum velocity—were captured [23,24]. They were tested during the preseason preparation period, when they engage in extensive physical training, wearing their regular running shoes for plyometric exercises. All trials were performed indoors on a tartan running track. Each trial began from a standing position, with a self-selected stance (one foot in front of the other). The jump exercises were performed in the following order: bounding for speed, followed by single leg jumps for speed on the right and left leg. Each participant completed a total of six trials: two for bounding and two for each leg in the single leg jump condition. The recovery time between attempts was 2–3 min.

2.3. Measurements

Contact and flight time were both measured with an IMU and an infrared optical contact mat (Ergotest Technology AS, Langesund, Norway). The IMUs (Ergotest Technology AS, Langesund, Norway) consisted of a wireless 3-axis accelerometer (±16 g, accuracy ±1.0%), a gyroscope (2000 deg/s, accuracy ±1.0%), and a magnetometer (±1300/2500 μT, accuracy ±5%), with a sampling frequency of 200 Hz and a weight of 20 g, fixed to the dorsal side of each foot with tape. The calculations of contact time and flight time with the IMU were determined by recognizing the velocity pattern of the plantar flexion/extension of both feet (Ergotest Technology AS, Langesund, Norway). The algorithm used to detect touch down and take off was originally developed for sprinting and running, where forefoot landings dominate. The infrared optical contact mat is composed of two 2.5 cm thick and 0.87 m long units ("master and slave") with a sampling frequency of 1000 Hz that were positioned at the start and end of the 20 m long test area. These units create a "mat" of infrared light approximately 5 mm above the floor, and breaks in this light (when the foot makes contact with the ground) are registered to record contact and flight times. Each participant's horizontal displacement was recorded with a laser gun (Noptel Oy, Oulu, Finland) directed towards their back during the tests. The laser gun continuously records distance over time using a CMP3 distance sensor (Noptel Oy, Oulu, Finland), sampling at 2.56 KHz. Horizontal speed was calculated using the distance over time during contact time with the laser. All recordings with the different sensors were synchronized and processed using Musclelab version 10.200.90.5095 (Ergotest Technology AS, Langesund, Norway).

Step-by-step analyses were performed on the first eight steps of each jump exercise, as this was the minimum number of steps completed by all participants. For each step, horizontal velocity and contact time were analysed using the IMU and IR mats. These parameters were automatically calculated by the Musclelab system for both measuring tools (Ergotest Technology AS, Langesund, Norway) and were earlier validated as very accurate in sprint step-by-step comparisons [14,16]. In addition, we examined the difference in total contact time, as well as the difference in the times when the foot hit the ground (touch down) and left the ground (take off). Flight times were not analysed, as they are the inverse of contact times when analysing absolute time measurements and, therefore, do not provide additional information about the differences between the two systems. Step length and frequency were also not analysed as these are directly related to the contact times and evt. differences would directly influence these parameters

2.4. Statistical Analysis

The data were tested for normality with the Shapiro–Wilk test and it was found that all data were normally distributed. To evaluate the validity between the two measurement devices and the two types of jumps, we performed two 2 (measuring device: IMU vs. IR mat/jump type: single leg jumps and bounding) × 8 (steps) ANOVA analyses with repeated measures. The first analysis compared the step velocities between the two jump types (single leg jumps and bounding) across eight steps. The second analysis evaluated the total contact times measured by the two tools (IR mat and IMU) over the same steps. Furthermore, a one-way ANOVA with repeated measures was performed on the differences in contact time between the two systems at touch down and take off for the eight steps. Where the sphericity assumption was violated, the Greenhouse–Geisser adjustments of the p-values were reported. The level of significance was set at $p < 0.05$. When significant differences were observed, post hoc comparisons were performed after Holm–Bonferoni correction. The effect size (Eta partial squared) was evaluated where $0.01 < \eta^2 < 0.06$ constitutes a small effect, $0.06 < \eta^2 < 0.14$ a medium one, and $\eta^2 > 0.14$ a large effect [25]. Bland–Altman analysis was used to examine systematic differences between the IR mat and IMU for touch down and time across different steps and velocities. Mean differences (systematic bias) and 95% confidence intervals were calculated and plotted as a function of speed. Regression analyses were performed to assess heteroscedasticity in the measurement differences. All data analyses were performed using JASP v. 0.17.3 (University of Amsterdam, Amsterdam, The Netherlands). Data were presented as means and standard deviations (SD).

3. Results

A significant effect of jump type ($F = 230$, $p < 0.001$, $\eta^2 = 0.45$), * steps ($F = 136$, $p < 0.001$, $\eta^2 = 0.36$), and jump type*steps interaction ($F = 25.7$, $p < 0.001$, $\eta^2 = 0.07$) was observed on step velocity, in which step velocity in during bounding increased significantly faster and longer than in single leg jumps and thereby resulted in significantly higher step velocity over the eight steps. In addition, one-way ANOVA with repeated measurements showed a significant effect of steps on speed for single leg jumps ($F = 15.655$, $p < 0.001$, $\eta^2 = 0.495$), and a significant effect was also observed for bounding ($F = 339.954$, $p < 0.001$, $\eta^2 = 0.955$), indicating a larger effect size for bounding (Figure 1).

When analysing the total contact times for each of the jump types between the two measuring tools, we found that significant effects on both types of jumps were caused by the measuring tool ($F \geq 415$, $p < 0.001$, $\eta^2 \geq 0.44$), steps ($F \geq 23.3$, $p < 0.001$, $\eta^2 \geq 0.43$), and measuring tool*steps interaction ($F \geq 2.7$, $p < 0.001$, $\eta^2 \leq 0.44$). Post hoc comparison reveals that contact times were consistently longer when measured with the IR-mat compared to the IMU. Furthermore, contact times decreased in both types of jumps, but the decrease over the steps was larger when measuring with the IMU compared to the IR-mat, especially in bounding. The analysis of total contact time confirmed this: for the IR-mat there was a significant effect ($F = 49.376$, $p < 0.001$, $\eta^2 = 0.755$), and for the IMU the effect was even greater ($F = 75.328$, $p < 0.001$, $\eta^2 = 0.825$). This indicates that both methods recorded a significant reduction in contact time over the steps, with the IMU recording greater changes overall. (Figure 2).

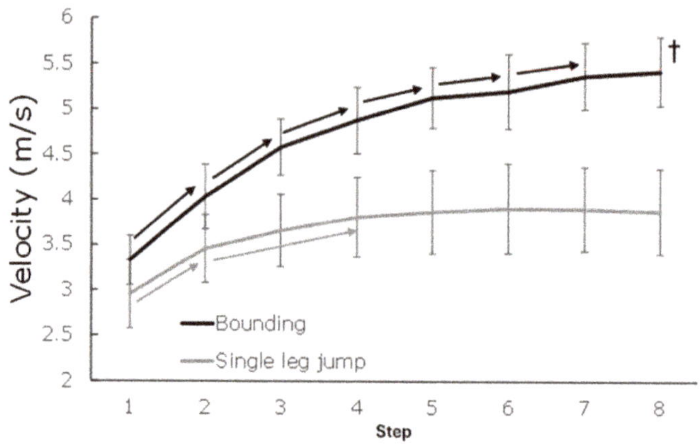

Figure 1. The mean (±SD) step velocity for bounding and single leg jumps measured with an IR-mat and a laser gun. † indicates a significant difference between the two types of jumps for all steps, → indicates a significant difference in velocity between this step and everything to the right of the arrow.

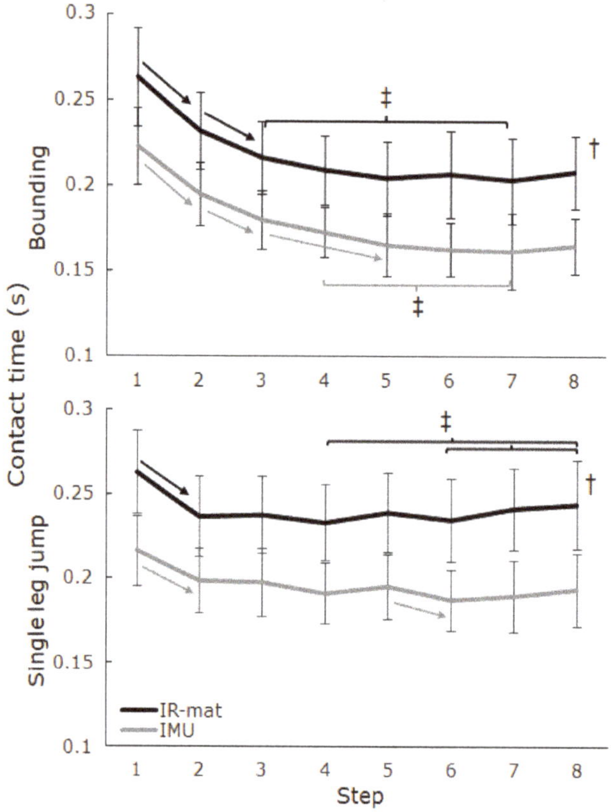

Figure 2. The mean (±SD) total contact time per step for bounding and single leg jumps measured with an IR-mat and an IMU. † indicates a significant difference between the IMU and the IR-mat for all steps, ‡ indicates a significant difference between these two steps for this measuring tool, and → indicates a significant difference between this step and everything to the right of the arrow.

The effect of velocity on the differences between the two systems was investigated using Bland–Altman plots, which evaluated the differences at touch down and take off. A systematic bias was identified: the IR mat consistently recorded an earlier start time for contact compared to the IMU at touch down, while it measured a later time at take off. This is reflected in the average bias: −0.018 s at touch down and 0.020 s at take off for bounding, and −0.024 s at touch down and 0.021 s at take off for single leg jumps (Figure 3). However, no significant effect of velocity ($r \leq 0.52$, $p > 0.05$) was found between these differences in times.

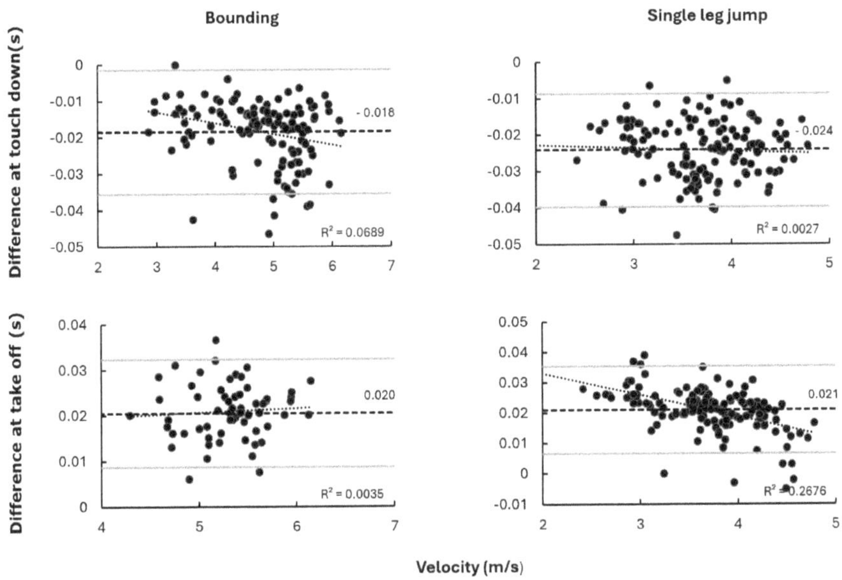

Figure 3. Bland–Altman plots showing the differences between the systems in regard to the time of touch down and take off (IR mat minus IMU, seconds) as a function of speed (m/s) for bounding and single leg jumps. Positive values indicate a longer time for the IR mat, negative values indicate a longer time for the IMU. The dashed lines show the average difference (systematic bias), while the solid lines represent 95% confidence intervals. Regression lines and R^2 values are included.

4. Discussion

The aim of this study was to compare step-by-step measurements from an IR-mat and an IMU during bounding and single leg jumps for speed. The main findings show significant differences between the two measurement systems in terms of total contact time, touch down, and take off.

Bounding jumps had the highest speed and the shortest contact time, in agreement with Mero and Komi [4]. Furthermore, we saw that the speed increased while contact time decreased throughout bounding, while for single leg jumps, we saw that this fluctuates more. This may be explained by the greater variability in landing patterns and the increased demands for balance and coordination in single leg jumps, which are inherently more complex exercises [12,15]. This study showed that the IR-mat consistently recorded a longer contact time compared to the IMU system. Previous studies have shown that IR-mats and IMUs are accurate in detecting contact time during sprint running, with only a minor difference of 0.003–0.004 s compared to a force plate [14,16]. From these studies, an algorithm to detect ground contact using an IMU was developed, in which an abrupt dip in angular velocity measured with the gyroscope together with a sudden acceleration spike were used to detect touch down and take off [14]. However, while sprinting a runner lands on their forefoot, while during bounding and single leg jumps they land on their heel [4]. This difference in landing pattern may contribute to the observed measurement differences, as the IMU pattern that identifies the touch down and take off through acceleration and angular velocity may perform differently with heel-based contact. Due to the mechanics of the heel-first landing, the back of the shoe, which has considerably more cushioning, will land first. Due to the shock absorption of the cushioning of the shoe, the IMU may detect the impact later than during sprint steps, in which athletes use spikes with much less cushioning and land on their mid/forefoot. This effect is further compounded by the location of the IMU, which is attached to the top of the shoe, which may result in slightly different IMU features.

The second factor contributing to the difference is the fact that the IR-mat starts to register foot contact approximately 5 mm above the ground. This setup means that it detects the foot slightly before touch down and stops recording a little after take off from the ground. Despite minor differences in the foot configuration at these points, the vertical speed at take off and touch down can be estimated using the following formula:

$$\frac{9.81^2 \times T}{2}$$

The average time in the air, measured by the IR-mat, was 0.198 s for bounding and 0.291 s for single leg jumps, which corresponds to vertical speeds at touch down and take off of 0.971 m/s and 1.427 m/s, respectively. The time it takes to move over 5 mm is calculated as follows:

$$t = \frac{d}{v}$$

For bounding this time is 0.005 s, while for single leg jumps it is 0.004 s. These differences explain only 15–25% of the variation between the two measurement methods, indicating that they cannot fully explain the observed deviations.

The third reason is the sample rate differences. The IR-mat records at 1000 Hz, while the IMU records at 200 Hz. In practice, this means that for each sample there is a maximum deviation of $\pm$0.005 s at touch down and take off.

Limitations and Future Studies

This study has some limitations that should be considered when interpreting the results. The participants were a homogeneous group of female football players with moderate experience in bounding and single leg jumps. Differences between the systems may arise when testing track and field athletes who are more experienced in these exercises and, therefore, may use different jumping techniques with higher velocities. Furthermore, the use of different footwear could lead to variations in landing and take off techniques, potentially influencing the results. This may limit the generalizability of the findings to other groups.

In addition, a gold standard, such as force plates combined with high-speed cameras, was not used, which prevents us from determining which system better aligns with the ground truth. Future studies should test both systems alongside force plates and high-speed cameras, with a broader range of athletes performing these types of jumps, to determine whether the observed differences are generalizable.

5. Conclusions

Our findings show significantly different measurements of contact times at touch down and take off between the IR-mat and the IMU in two different horizontal jump types. The IR-mat consistently recorded longer contact times by starting the measurements a bit earlier at touch down and ending slightly later than the IMU system at take off. The main reason for these differences is likely the fact that the algorithm for detecting touch down and take off was designed for sprint running and landing on the forefoot with sprint spikes, whereas the participants in this study landed on their heels and wore soft running shoes. Furthermore, the differences in sampling rate and the height of detection of the IR-mat may partially contribute to the differences in these parameters.

In practical terms, when choosing a measurement system for horizontal jumps, it is important to be cautious when comparing spatiotemporal parameters measured with different systems. IR-mats and IMUs differ in how they detect events like touch down and take off, with IR-mats recording these events directly and IMUs relying on algorithms. This can lead to variations, especially during heel-based landings in soft running shoes, highlighting the need for careful interpretation of results.

Author Contributions: Conceptualization: B.J. and R.v.d.T.; methodology: B.J. and R.v.d.T.; data collection and curation: B.J.; formal analysis: B.J. and R.v.d.T.; writing—original draft preparation: B.J.; writing—review and editing: R.v.d.T. and J.N.; visualization: B.J.; supervision: R.v.d.T. and J.N.; project administration: B.J. All authors have read and agreed to the published version of the manuscript.

Funding: This research received no external funding.

Institutional Review Board Statement: This study was conducted in accordance with the Declaration of Helsinki and approved by the Norwegian Center for Research Data approved 12 November 2021 (project number: 225529).

Informed Consent Statement: Informed consent was obtained from all subjects involved in the study.

Data Availability Statement: The data presented in this study are available on request from the corresponding author. The data are not publicly available due to the national laws on privacy of the Norwegian government.

Conflicts of Interest: The authors declare no conflicts of interest.

References

1. Montoro-Bombú, R.; Miranda-Oliveira, P.; Valamatos, M.J.; João, F.; Buurke, T.J.; Santos, A.C.; Rama, L. Spatiotemporal variables comparison between drop jump and horizontal drop jump in elite jumpers and sprinters. *PeerJ* **2024**, *12*, e17026. [CrossRef]
2. Wang, Y.; Mei, Q.; Jiang, H.; Hollander, K.; Van den Berghe, P.; Fernandez, J.; Gu, Y. The Biomechanical Influence of Step Width on Typical Locomotor Activities: A Systematic Review. *Sports Med. Open* **2024**, *10*, 83. [CrossRef]
3. Möhler, F.; Fadillioglu, C.; Stein, T. Fatigue-related changes in spatiotemporal parameters, joint kinematics and leg stiffness in expert runners during a middle-distance run. *Front. Sports Act. Liv.* **2021**, *3*, 634258. [CrossRef]
4. Mero, A.; Komi, P.V. EMG, force, and power analysis of sprint-specific strength exercises. *J. Appl. Biomech.* **1994**, *10*, 1–13. [CrossRef]
5. Bobbert, M.F. Why do people jump the way they do? *Exerc. Sport Sci. Rev.* **2001**, *29*, 95–102. [CrossRef] [PubMed]
6. Luebbers, P.E.; Potteiger, J.A.; Hulver, M.W.; Thyfault, J.P.; Carper, M.J.; Lockwood, R.H. Effects of plyometric training and recovery on vertical jump performance and anaerobic power. *J. Strength Cond. Res.* **2003**, *17*, 704–709. [PubMed]
7. Mero, A.; Komi, P.V. Effects of supramaximal velocity on biomechanical variables in sprinting. *J. Appl. Biomech.* **1985**, *1*, 240–252. [CrossRef]
8. Miyashiro, K.; Nagahara, R.; Yamamoto, K.; Nishijima, T. Kinematics of maximal speed sprinting with different running speed, leg length, and step characteristics. *Front. Sports Act. Liv.* **2019**, *1*, 37. [CrossRef]

9. Ettema, G.; McGhie, D.; Danielsen, J.; Sandbakk, Ø.; Haugen, T. On the existence of step-to-step breakpoint transitions i accelerated sprinting. *PLoS ONE* **2016**, *11*, e0159701. [CrossRef] [PubMed]
10. van den Tillaar, R. Comparison of development of step-kinematics of assisted 60 m sprints with different pulling forces betwee experienced male and female sprinters. *PLoS ONE* **2021**, *16*, e0255302. [CrossRef] [PubMed]
11. Blauberger, P.; Horsch, A.; Lames, M. Detection of ground contact times with inertial sensors in elite 100-m sprints unde competitive field conditions. *Sensors* **2021**, *21*, 7331. [CrossRef] [PubMed]
12. Schmidt, M.; Rheinländer, C.; Nolte, K.F.; Wille, S.; Wehn, N.; Jaitner, T. IMU-based determination of stance duration durin sprinting. *Proc. Engin.* **2016**, *147*, 747–752. [CrossRef]
13. Zeng, Z.; Liu, Y.; Hu, X.; Tang, M.; Wang, L. Validity and reliability of inertial measurement units on lower extremity kinematic during running: A systematic review and meta-analysis. *Sports Med. Open* **2022**, *8*, 86. [CrossRef]
14. van den Tillaar, R. Comparison of step-by-step kinematics of elite sprinters' unresisted and resisted 10-m sprints measured wit Optojump or Musclelab. *J. Strength Cond. Res.* **2021**, *35*, 1419–1424. [CrossRef]
15. Healy, R.; Kenny, I.C.; Harrison, A.J. Assessing reactive strength measures in jumping and hopping using the Optojump™ systen *J. Hum. Kinet.* **2016**, *54*, 23. [CrossRef] [PubMed]
16. van den Tillaar, R.; Nagahara, R.; Gleadhill, S.; Jiménez-Reyes, P. Step-to-step kinematic validation between an Inertial Measure ment Unit (IMU) 3D system, a combined Laser+IMU system and force plates during a 50 M sprint in a cohort of sprinters. *Sensor* **2021**, *21*, 6560. [CrossRef] [PubMed]
17. García-Pinillos, F.; Jaén-Carrillo, D.; Hermoso, V.S.; Román, P.L.; Delgado, P.; Martinez, C.; Carton, A.; Seruendo, L.R. Agreemen Between Spatiotemporal Gait Parameters Measured by a Markerless Motion Capture System and Two Reference Systems—A Treadmill-Based Photoelectric Cell and High-Speed Video Analyses: Comparative Study. *JMIR mHealth uHealth* **2020**, *8*, e1949 [CrossRef] [PubMed]
18. Ding, Y.; Xiong, Z.; Li, W.; Cao, Z.; Wang, Z. Pedestrian navigation system with trinal-IMUs for drastic motions. *Sensors* **202** *20*, 5570. [CrossRef] [PubMed]
19. de Ruiter, C.J.; van Dieën, J.H. Stride and step length obtained with inertial measurement units during maximal sprint acceleratio *Sports* **2019**, *7*, 202. [CrossRef]
20. Weyand, P.G.; Sternlight, D.B.; Bellizzi, M.J.; Wright, S. Faster top running speeds are achieved with greater ground forces no more rapid leg movements. *J. Appl Physiol.* **2000**, *89*, 1991–1999. [CrossRef] [PubMed]
21. Swearingen, J.; Lawrence, E.; Stevens, J.; Jackson, C.; Waggy, C.; Davis, D.S. Correlation of single leg vertical jump, single leg ho for distance, and single leg hop for time. *Phys. Ther. Sport* **2011**, *12*, 194–198. [CrossRef] [PubMed]
22. FIFA. FIFA 11+ Manual. Available online: https://docslib.org/doc/876010/fifa-11-manual-pdf (accessed on 11 July 2024).
23. Kariyama, Y.; Hobara, H.; Zushi, K. The effect of increasing jump steps on stance leg joint kinetics in bounding. *Int. J. Sports Me* **2018**, *39*, 661–667. [CrossRef] [PubMed]
24. Kariyama, Y.; Hobara, H.; Zushi, K. Differences in take-off leg kinetics between horizontal and vertical single-leg rebound jump *Sports Biomech.* **2017**, *16*, 187–200. [CrossRef] [PubMed]
25. Cohen, J. *Statistical Power Analysis for the Behavioral Sciences*; Lawrence Erlbaum Associates: Hillsdale, NJ, USA, 1988; p. 174.

biomechanics

MDPI

Article

Assessing the Contribution of Arm Swing to Countermovement Jump Height Using Three Different Measurement Methods in Physically Active Men

Daichi Yamashita *, Frederick James Henderson and Yuko Ishida

Japan Institute of Sports Sciences, 3-15-1, Nishigaoka, Kita-ku, Tokyo 115-0056, Japan; frederick.henderson@jpnsport.go.jp (F.J.H.)

* Correspondence: daichi.yamashita@jpnsport.go.jp

Abstract: Background/Objectives: This study evaluated the reliability and validity of three methods to measure jump height during countermovement jumps performed with (CMJ_{AS}) and without (CMJ_{NAS}) arm swing: (1) an impulse–momentum method with force platforms (FP_{imp}), (2) a flight time method with force platforms (FP_{time}), and (3) an inertial measurement unit (PUSH Band 2.0; PUSH2). Methods: Eighteen physically active men performed CMJ_{AS} and CMJ_{NAS} on force platforms while wearing PUSH2 over two days. Besides jump height, we computed intraclass correlation coefficients (ICC) and the absolute and relative increases in jump height due to arm swing, compared to CMJ_{NAS}. Results: The reliability of intra-session, inter-session, and concurrent measures were good to excellent (intra-session $ICC_{2,1}$ = 0.957–0.979, inter-session $ICC_{2,1}$ = 0.806–0.990, concurrent $ICC_{3,1}$ = 0.940–0.973) for CMJ_{AS} and CMJ_{NAS} heights, in all three methods. The three methods showed high to very high reliability for both the absolute and relative indices of arm swing contribution ($ICC_{2,1}$ = 0.649–0.812). FP_{time} significantly overestimated CMJ_{NAS} height relative to FP_{imp} ($p < 0.01$). The absolute index of arm swing contribution was similar in FP_{imp} and FP_{time} ($p = 0.38$) but higher in PUSH2 ($p < 0.01$), indicating that arm swing amplified overestimation. Conclusions: All three methods demonstrated high reliability for jump height measurements, but FP_{time} and PUSH2 misestimated jump height depending on jump modalities. Caution is advised when assessing the absolute and relative contribution of arm swing, because errors in CMJ_{NAS} and CMJ_{AS} height measurements can affect these values and their interpretation.

Keywords: inertial measurement unit; arm contribution index; athlete monitoring; training; vertical jump

1. Introduction

Jump performance evaluation is an important training tool in many sports, e.g., American football [1] and basketball [2]. For example, routine countermovement jump (CMJ) testing helps practitioners prescribe exercises and manage fatigue in athletes [3–5]. While various technologies and methods have emerged to measure CMJ performance, the validity and reliability of these measurement methods must be rigorously established to ensure diagnostic accuracy. Yet, this information is often lacking.

CMJ height is commonly measured with or without arm swing [6–9], with arm swing possibly enhancing CMJ height [6,10]. For example, previous research suggests that athletes can improve jump height by over 20% with arm swing compared to without arm swing [6,7,11], whereas non-athletes show more modest improvements [12]. Therefore, not only are reliable and valid measures of CMJ height crucial for practitioners and researchers, but understanding whether arm swing alters the reliability and validity of measurements is also important in evaluating athletes.

Different methods have different merits and demerits. The gold standard for measuring jump height involves recording ground reaction forces using force platforms (FPs) to derive the impulse (FP_{imp}), utilizing the impulse–momentum relationship [13,14]. However, while extensively validated, implementing FP_{imp} can be challenging due to the hardware costs, limited portability, and complex calculation processes. Alternatively, jump height can be derived from the flight time (FP_{time}) using not only a FP but also simpler, more affordable technology like a contact mat, photoelectric cell, or smartphone application [15]. Although the flight time method may overestimate jump height compared with FP_{imp}, it remains widely used because the overestimation is considered reasonable [14,16]. Another method involves the use of commercially available inertial measurement units (IMUs) consisting of wireless accelerometers and gyroscopes. While all IMUs rely on similar technology, manufacturers often use proprietary computing algorithms generally treated as a "black box" [17,18]. One such IMU popular among strength and conditioning coaches is PUSH Band 2.0 (PUSH2; Push Inc., Toronto, Canada). To the authors' knowledge, few studies have examined the validity and reliability of PUSH2, showing reliable measures but overestimating CMJ without arm swing (CMJ_{NAS}) height compared to FP_{imp} [19] or the flight time method using FPs [20] and optical sensors [21]. However, although the PUSH2 approach is an attractive approach due to its lower cost, whether its measures are robust across CMJ modalities, i.e., CMJ with arm swing (CMJ_{AS}) or CMJ_{NAS}, has yet to be established.

In particular, it remains unclear whether the arm swing amplifies these overestimations. For example, CMJ height overestimation may affect the calculation of arm contribution—the difference in CMJ height between CMJ_{NAS} and CMJ_{AS} or the ratio of CMJ_{AS} and CMJ_{NAS} [6,7,11]—thus yielding an erroneous diagnostic. Indeed, if both CMJ_{AS} and CMJ_{NAS} heights are overestimated by the same amount (e.g., 4 cm), the relative arm contribution would appear smaller with PUSH2 or FP_{time} than with FP_{imp} because the denominator (CMJ_{NAS} height) increases. Consequently, investigating how arm swing impacts IMU system reliability and validity compared to FP_{time} and FP_{imp} would be valuable for practitioners to select the appropriate measuring tool.

Therefore, the purpose of this study was (1) to investigate the impact of the FP_{imp}, FP_{time}, and PUSH2 methods on the reliability and validity of CMJ_{NAS} and CMJ_{AS} height measurements, and (2) to determine how the observed misestimation influences the computation of arm contribution indices. We hypothesized that the degree of overestimation in jump height calculated using the IMU compared to the FP_{imp} method would be greater than using the FP_{time} method, whereas the degree of overestimation would be comparable between CMJ_{NAS} and CMJ_{AS}, regardless of the method used. Such outcomes would suggest that the evaluation of arm contributions can vary depending on the method employed.

2. Materials and Methods

2.1. Participants

Eighteen physically active men (age: 34.4 ± 5.4 years; height: 171.8 ± 6.1 cm; body mass: 74.1 ± 8.6 kg), free from any injury affecting CMJ performance, provided written informed consent to participate in this study. This study was approved by the Institutional Ethics Committee of the Japanese Institute of Sports Sciences (No. 2019-062, approved on 10 February 2020) in accordance with the Declaration of Helsinki. Power analysis using G*Power software (version 3.1.9.2, Heinrich-Heine-Universität Düsseldorf, Düsseldorf, Germany) showed that to achieve a power of 80% at an alpha level of 5%, at least eighteen participants were needed to achieve a medium effect size of partial $\eta_p^2 = 0.06$ ($f = 0.253$) [22] in a within-subject repeated-measures 3 × 2 analysis of variance (ANOVA) (1 group, 6 measurements, correlation among measures = 0.5, non-sphericity correction = 1).

2.2. Procedures

To evaluate inter-day reliability, all participants completed two testing sessions 3–7 days apart, as used in previous research [23–25], to allow for neuromuscular recovery, limit possible fatigue, and minimize any potential learning effect. Testing was conducted around the same time of the day (within a 2 h window) to minimize circadian influences [26]. The participants were also asked to maintain their normal routines and avoid strenuous exercise 48 h before each test and during the interval between sessions.

The participants first completed a standardized warm-up consisting of jogging, dynamic stretching, and jumping. PUSH2 was then attached to the participant's waist using the manufacturer-provided waist belt. Finally, the participants performed three CMJ_{NAS} and three CMJ_{AS} with each foot on a different FP (1200 Hz, 0.6 m × 0.4 m, type 9281E; Kistler, Winterthur, Switzerland). For CMJ_{NAS}, the participants were instructed to jump as high as possible using a rapid countermovement while keeping their hands on their hips. For CMJ_{AS}, they were instructed to jump as high as possible using a rapid countermovement while swinging their arms. For both jump types, the participants were instructed to initiate landing with their legs straight to limit the influence of posture on flight time. In addition, for CMJ_{AS}, they were asked to land with their arms raised to approximately the same height as at take-off to minimize potential overestimation of CMJ_{AS} height (Figure 1) [14]. A minimum of 30 s of rest was given between each jump trial and 180 s of rest between each jump modality. One experimenter watched each trial and, if instructions were not respected (e.g., leg tucking upon landing), the trial was repeated 30 s later.

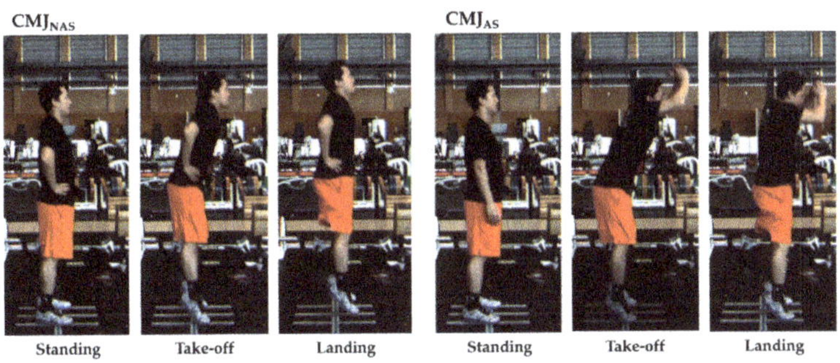

Figure 1. Illustration of the countermovement jump tasks: without arm swing (CMJ_{NAS}; **left**) and with arm swing (CMJ_{AS}; **right**).

Unfiltered left- and right-side vertical ground reaction forces (GRF_v) were then summed for analysis, because low-pass filtering can cause substantial underestimation of jump height [27]. Jump height obtained from PUSH2 (sampled at 1000 Hz) was sent via Bluetooth to an iPad Pro (Apple Inc., Cupertino, CA, USA) using the manufacturer's software (PUSH Pro, v7.7.0). Body weight was calculated from the 0.5 s (600 data points) moving average of GRFv, shifted by one frame, displaying the smallest standard deviation (SD) during quiet standing. The start of the motion was identified as the moment when GRF_v deviated from body weight by 1% [14], and take-off was defined as the first point when GRFv fell below 10 N (Figure 2) [28]. To estimate the vertical velocity at take-off (V_{to}), GRF_v was integrated from the start of the motion until take-off by trapezoid rule integration [14]. Jump height from FP_{imp} was then calculated as

$$Jump\ height\ from\ FP_{imp} = \frac{1}{2g}V_{to}^2, \quad (1)$$

where g represents the gravitational acceleration (9.81 m/s^2). Jump height from FP_{time} was calculated as [20]

$$Jump\ height\ from\ FP_{time} = \frac{1}{8}gT_{flight}^2, \quad (2)$$

where T_{flight} represents the flight time from take-off to landing, defined as the first and second time that GRF_v crosses the 10 N threshold [28]. The results from FP_{imp} were the reference measurements [19].

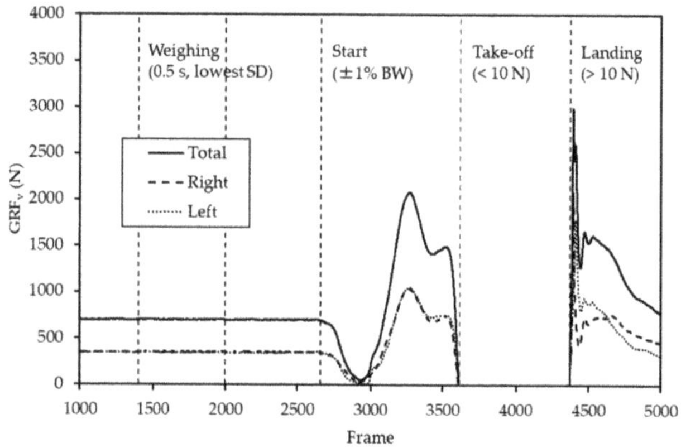

Figure 2. Force–time curves of vertical ground reaction forces (GRFv) from the left and right legs during a countermovement jump, showing key events and definitions. SD = standard deviation; BW = body weight.

The arm contribution indices (AI) were defined as previously reported [7,8]. The absolute AI (AI_{abs}) was thus the difference in CMJ height with and without arm swing:

$$AI_{abs}(cm) = CMJ_{AS} - CMJ_{NAS}. \quad (3)$$

The relative AI (AI_{rel}) was AI_{abs} relative to CMJ_{NAS}:

$$AI_{rel}(\%) = \frac{AI_{abs}}{CMJ_{NAS}} \times 100. \quad (4)$$

2.3. Statistical Analysis

The height of each participant's three CMJ_{NAS} or CMJ_{AS} trials was used to compute intra-session and concurrent reliability, along with Bland–Altman analyses. The average of the three trials was used for the other analyses, including ANOVAs, assessments of inter-day test–retest reliability, and concurrent validity. Means and SDs were calculated after confirming the data's normal distribution using Kolmogorov–Smirnov statistics ($p \geq 0.05$).

The intra-session reliability for CMJ_{NAS} and CMJ_{AS} height from the FP_{imp}, FP_{time}, and PUSH2 methods was determined by calculating the coefficient of variation (CV) and the intraclass correlation coefficient ($ICC_{2,1}$; two-way mixed-effects model with absolute agreement) with a 95% confidence interval (CI) [29–31]. In addition, the test–retest inter-day reliability of all variables between Day 1 and 2 was assessed using $ICC_{2,1}$ (i.e., two-way random-effects model with absolute agreement) with a 95% CI [31].

A two-way repeated-measures ANOVA with Bonferroni post hoc paired t-tests was used to evaluate the effect of the two jump modalities (CMJ_{AS} and CMJ_{NAS}) and the three methods on CMJ height. One-way repeated-measures ANOVAs with Bonferroni post hoc paired t-tests were used to test the differences in AI_{abs} and AI_{rel} across the three methods. The Greenhouse–Geisser correction was applied if the Mauchly sphericity test revealed that the assumption of sphericity was violated. Partial eta-squared values (η_p^2) were reported as a measure of effect size: small (0.01–0.06), medium (0.06–0.14), and large (>0.14) [22].

The concurrent validity among the three methods was assessed using Pearson's correlation coefficients (r). To further evaluate agreement, Bland–Altman plots were analyzed to quantify the systematic bias and 95% limits of agreement (LoA) among the three methods [30,32]. The 95% LoA was corrected for repeated measures [33]. Proportional bias was identified when the coefficient of determination (R^2) was greater than 0.1 [34]. In addition, concurrent reliability among the three methods was assessed using $ICC_{3,1}$ (i.e., two-way mixed-effects model for consistency) with a 95% CI [31].

For ICC and r interpretations, we used the following criteria: poor (<0.50), moderate (0.50–0.75), good (0.75–0.90), and excellent (>0.90) [31]. For CV, we used the following criteria: poor (>10%), moderate (5–10%), and good (<5%) [35,36]. Statistical significance was set at $p < 0.05$. Analyses were performed using SPSS 19.0 (IBM Corp., Chicago, IL, USA) and Microsoft Excel (Microsoft Corporation, Redmond, WA, USA).

3. Results

Intra-session reliability analysis showed good reliability based on CV (1.56–2.48%) and excellent reliability based on $ICC_{2,1}$ (0.96–0.98) for all variables (Table 1). The inter-day test–retest reliability analysis showed excellent reliability for CMJ_{NAS} and CMJ_{AS} height in each method ($ICC_{2,1}$ = 0.940–0.973), but moderate to excellent reliability for AI_{abs} and AI_{rel} in all three methods ($ICC_{2,1}$ = 0.62–0.78) (Table 2).

Table 1. Intra-session reliability of the CMJ height from FP_{imp}, FP_{time}, and PUSH2.

		Day 1		Day 2	
		Mean CV (95% CI)	$ICC_{2,1}$ (95% CI)	Mean CV (95% CI)	$ICC_{2,1}$ (95% CI)
CMJ_{NAS}	FP_{imp}	1.89 (1.26, 2.52)	0.98 (0.96, 0.99)	2.21 (1.45, 2.98)	0.97 (0.93, 0.99)
	FP_{time}	1.91 (1.31, 2.52)	0.97 (0.94, 0.99)	2.48 (1.63, 3.33)	0.95 (0.89, 0.98)
	PUSH2	1.98 (1.40, 2.57)	0.98 (0.95, 0.99)	2.37 (1.38, 3.35)	0.93 (0.86, 0.97)
CMJ_{AS}	FP_{imp}	1.68 (1.15, 2.21)	0.97 (0.94, 0.99)	1.56 (0.97, 2.15)	0.98 (0.95, 0.99)
	FP_{time}	1.99 (1.33, 2.65)	0.96 (0.92, 0.99)	2.06 (1.29, 2.83)	0.96 (0.90, 0.98)
	PUSH2	1.60 (1.21, 2.84)	0.96 (0.91, 0.98)	1.69 (1.22, 2.17)	0.96 (0.91, 0.98)

FP_{imp} = impulse–momentum method using force platforms; FP_{time} = flight time method using force platforms; PUSH2 = PUSH Band 2.0; CMJ_{NAS} = countermovement jump without arm swing; CMJ_{AS} = countermovement jump with arm swing; CV = coefficient of variation; CI = confidence interval; ICC = intraclass correlation coefficient.

The two-way repeated-measures ANOVA revealed a significant interaction between jump modalities and methods in CMJ height (F [1.50, 25.47] = 14.22, $p < 0.001$, $\eta_p{}^2 = 0.455$, large effect size) (Table 2). Regarding jump modalities, post hoc paired t-tests with Bonferroni corrections revealed that CMJ_{AS} height was significantly greater than CMJ_{NAS} height across all three measurement methods ($p < 0.01$). Furthermore, regarding measurement methods, both the CMJ_{NAS} height and the CMJ_{AS} height were greater for PUSH2 than FP_{time} and greater for FP_{time} than FP_{imp} ($p < 0.01$).

One-way repeated-measures ANOVA revealed a significant main effect of measurement methods on AI_{abs} (Day 1: F [2, 34] = 14.21, $p < 0.001$, $\eta_p{}^2 = 0.455$; Day 2: F [2, 34] = 14.03, $p < 0.001$, $\eta_p{}^2 = 0.452$) and AI_{rel} (Day 1: F [1.50, 25.55] = 9.11, $p < 0.01$, $\eta_p{}^2 = 0.349$; Day 2: F [2, 34] = 10.71, $p < 0.01$, $\eta_p{}^2 = 0.386$), indicating large effect sizes across both days (Table 3). On both days, AI_{abs} was significantly higher when using PUSH2 compared to FP_{imp} ($p < 0.01$) and FP_{time} ($p < 0.01$). However, there was no significant difference in AI_{abs} between FP_{time} and FP_{imp} (Day 1: $p = 0.384$; Day 2; $p = 0.60$) (Table 3). This greater difference in AI_{abs} from PUSH2 explains the significant interaction between jump modality and measurement method. For AI_{rel}, values were significantly lower when using FP_{time} compared to PUSH2 ($p < 0.01$) and FP_{imp} ($p = 0.04$) on Day 1. On Day 2, AI_{rel} from FP_{time} remained significantly lower than PUSH2 ($p < 0.01$), while the difference in AI_{rel} between FP_{time} and FP_{imp} was not statistically significant ($p = 0.23$).

Table 2. Comparison of CMJ heights measured using three methods across two jump modalities, with corresponding inter-day reliability.

		Mean ± SD (Day 1)	Mean ± SD (Day 2)	$ICC_{2,1}$ (95% CI)
CMJ_{NAS} (cm)				
	FP_{imp}	40.04 ± 6.36	39.86 ± 6.19	0.97 (0.93, 0.99)
	FP_{time}	42.27 ± 6.26 †	41.86 ± 5.87 †	0.96 (0.90, 0.99)
	PUSH2	44.03 ± 6.46 †‡	43.24 ± 5.41 †‡	0.96 (0.89, 0.98)
CMJ_{AS} (cm)				
	FP_{imp}	44.90 ± 6.07 *	44.41 ±6.18 *	0.96 (0.91, 0.99)
	FP_{time}	46.67 ± 5.55 † *	45.92 ± 6.17 † *	0.95 (0.86, 0.98)
	PUSH2	50.68 ± 7.72 †‡ *	49.42 ± 6.87 †‡ *	0.94 (0.81, 0.98)
AI_{abs} (cm)				
	FP_{imp}	4.86 ± 2.18	4.54 ± 2.16	0.77 (0.48, 0.91)
	FP_{time}	4.40 ± 2.68	4.05 ± 2.13	0.65 (0.28, 0.85)
	PUSH2	6.65 ± 3.50 †‡	6.18 ± 2.86 †‡	0.81 (0.57, 0.93)
AI_{rel} (%)				
	FP_{imp}	12.72 ± 6.26 ‡	11.77 ± 5.81	0.74 (0.44, 0.89)
	FP_{time}	10.75 ± 6.37	9.91 ± 5.50	0.62 (0.23, 0.84)
	PUSH2	15.39 ± 8.29 ‡	14.37 ± 6.54 ‡	0.78 (0.50, 0.91)

FP_{imp} = impulse–momentum method using force platforms; FP_{time} = flight time method using force platforms; PUSH2 = PUSH Band 2.0; CMJ_{NAS} = countermovement jump without arm swing; CMJ_{AS} = countermovement jump with arm swing; AI_{abs} = absolute arm contribution index; AI_{rel} = relative arm contribution index; CV = coefficient of variation; CI = confidence interval; ICC = intraclass correlation coefficient. * Significantly higher than CMJ_{NAS} ($p < 0.05$). † Significantly higher than FP_{imp} ($p < 0.05$). ‡ Significantly higher than FP_{time} ($p < 0.05$).

Table 3. Systematic bias, proportional bias, and concurrent reliability of CMJ height measurements.

	Day 1			Day 2		
	Systematic Bias (cm) (95% LoA)	R^2	$ICC_{3,1}$ (95% CI)	Systematic Bias (cm) (95% LoA)	R^2	$ICC_{3,1}$ (95% CI)
CMJ_{NAS}						
FP_{time} vs. FP_{imp}	2.23 (−1.83, 6.29)	<0.01	0.95 (0.87, 0.98)	2.00 (−2.24, 6.24)	0.02	0.95 (0.86, 0.98)
PUSH2 vs. FP_{imp}	3.99 (−0.73, 8.72)	<0.01	0.94 (0.84, 0.98)	3.38 (−0.97, 7.73)	0.17	0.94 (0.84, 0.98)
PUSH2 vs. FP_{time}	1.76 (−1.89, 5.42)	<0.01	0.96 (0.90, 0.99)	1.38 (−3.69, 6.46)	0.12	0.96 (0.89, 0.98)
CMJ_{AS}						
FP_{time} vs. FP_{imp}	1.77 (−3.16, 6.70)	0.04	0.93 (0.83, 0.98)	1.51 (−3.31, 6.33)	<0.01	0.93 (0.84, 0.98)
PUSH2 vs. FP_{imp}	5.78 (−0.16, 11.82)	0.15	0.91 (0.77, 0.97)	5.01 (−1.47, 11.50)	<0.01	0.88 (0.70, 0.95)
PUSH2 vs. FP_{time}	4.01 (−1.27, 9.29)	0.06	0.93 (0.83, 0.97)	3.50 (−1.03, 8.04)	<0.01	0.94 (0.86, 0.98)

FP_{imp} = impulse–momentum method using force platforms; FP_{time} = flight time method using force platforms; PUSH2 = PUSH Band 2.0; CMJ_{NAS} = countermovement jump without arm swing; CMJ_{AS} = countermovement jump with arm swing; LoA = limit of agreement; CI = confidence interval. R^2 = coefficient of determination. The 95% LoA was corrected for repeated measures [33].

For concurrent validity, Pearson's correlation coefficients revealed excellent agreement between the methods ($r = 0.92$–0.96) (Figure 3). Concurrent reliability, accounting for within-subject variation across repeated trials, was also excellent between methods (ICC (3,1) = 0.88–0.96) (Table 3).

The systematic and proportional biases between measurement devices, assessed using Bland–Altman plots, are summarized in Table 3. For CMJ_{AS} height on Day 1, a proportional bias was observed between FP_{imp} and PUSH2 ($R^2 = 0.15$). On Day 2, proportional bias was found for CMJ_{NAS} height between FP_{imp} and PUSH2 ($R^2 = 0.17$) and FP_{time} and PUSH2 ($R^2 = 0.12$).

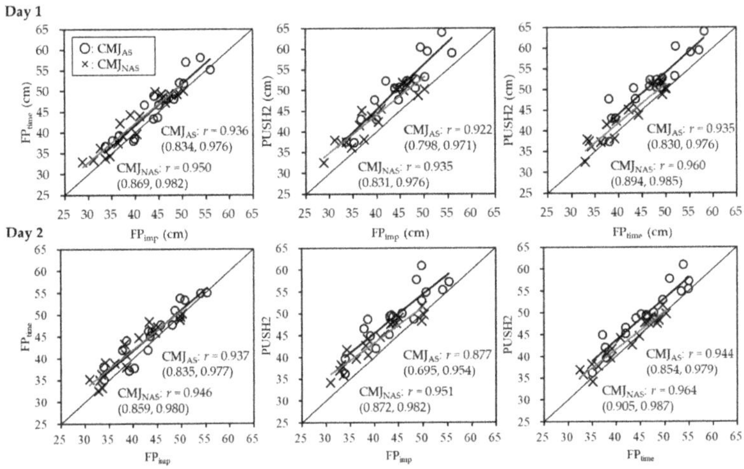

Figure 3. Correlations between CMJ height for FP_{imp}, FP_{time}, and PUSH2.

4. Discussion

This study evaluated the reliability and validity of force platform-derived impulse (FP_{imp}), force platform-derived flight time (FP_{time}), and an inertial measurement unit (PUSH2) in assessing CMJ height, both with and without arm swing, and subsequently examined the implications of these methods on arm contribution indices. The main findings were that FP_{time} and PUSH2 can be considered reliable methods for assessing CMJ_{NAS} and CMJ_{AS} heights, but both overestimated CMJ_{AS} and CMJ_{NAS} heights. Among the three methods, PUSH2 overestimated AI_{abs} and AI_{rel} the most and was less reliable between sessions.

The intra-session and inter-day reliability of the CMJ_{NAS} height for all three methods was good to excellent (ICC > 0.75) [31]. This result aligns with previous studies using FP_{imp}, FP_{time}, and PUSH2 [19–21]. In addition, for all three methods, the intra-session and inter-day reliability of CMJ_{AS} height in the present study was acceptable. Consequently, all these methods offered reliable and valid evaluations of jump height.

Regarding CMJ height, there was an interaction between jump modalities and measurement methods. CMJ heights were consistently ranked, from the highest to the lowest, PUSH2-FP_{time}-FP_{imp} for both CMJ_{AS} and CMJ_{NAS}. Critically, AI_{abs} using PUSH2 was significantly greater than FP_{imp} and FP_{time}. Hence, the arm swing amplified the intrinsic overestimation of jump height from PUSH2, but not FP_{time}. Although the coefficients of determination were not large and the specific jump modality showing a bias varied by day, given the proportional bias observed between FP_{imp} and PUSH2, PUSH2 may differentially affect CMJ height measurements based on jump performance level. However, PUSH Inc. has not disclosed their algorithms, making it uncertain whether the algorithms for the two CMJ modalities are identical. This is problematic because the computation algorithm, which includes data filtering, can have a sizeable effect on the measurement [37]. Still, we can speculate that the kinematics of the IMU device fixed on the waist level would vary between CMJ_{AS} and CMJ_{NAS} among participants (i.e., owing to trunk length, angle, and angular velocity). For example, the trunk segment at take-off may be more forward-tilted in untrained participants [38] but more upright in volleyball players [39] during CMJ_{AS} than during CMJ_{NAS}. Consequently, kinematic differences in the trunk might lead to a more pronounced overestimation of jump height from PUSH2 attached to the participants' waist when arm swing is involved.

In this study, arm swing improved jump height by 4.86 cm (12.72%) for FP_{imp}, 4.40 cm (10.75%) for FP_{time}, and 6.65 cm (15.39%) for PUSH2. Previous studies have suggested that arm swing enhances the effect of proximal-to-distal sequencing during jumping [10], and the contribution of CMJ height influences proper jump techniques in sports such as volleyball [11]. However, our results revealed that the metrics used to evaluate arm contribution (i.e., AI_{abs} and AI_{rel}) differed across methods. PUSH2 overestimated AI_{rel} because its overestimation of CMJ_{AS} height was greater than that of CMJ_{NAS} height, but FP_{time} underestimated AI_{rel} by similarly overestimating CMJ_{NAS} and CMJ_{AS}, which appear in the numerator and denominator of Equation (4). A previous study speculated that the overestimation by FP_{time} in CMJ_{AS} would be greater than that in CMJ_{NAS} because the height of the arm at take-off tends to be higher than that at landing [14]. In contrast, participants in this study were instructed to ensure that their arm height at landing matched that at take-off. This suggests that specific instructions regarding arm swing at landing could effectively mitigate the potential systematic bias of the CMJ_{AS} height from FP_{time}.

Regarding inter-session reliability, AI_{abs} and AI_{rel} displayed lower ICCs than CMJ_{AS} and CMJ_{NAS}, although CMJ_{AS} and CMJ_{NAS} heights from the three methods demonstrated acceptable inter-day reliability. Specifically, while the reliability for FP_{time} was below the acceptable threshold, the reliability of the other methods slightly exceeded this threshold. It is crucial to consider that indices derived from two numbers typically exhibit lower reliability, owing to the inherent variability of both numbers [40]. Thus, caution is advised when using such indices. Additionally, it is worth highlighting that the participants in this study were neither competitive athletes nor engaged in sports involving arm-swing techniques. AI_{rel} from FP_{imp} (12–13%) was considerably lower than the 24% in collegiate male and female basketball players reported in Heishman et al. [6] or the 38% in elite male volleyball players reported in Vaverka et al. [11]. Consequently, our findings may not apply to populations with pronounced arm contribution, such as basketball and volleyball athletes used to jumping with an arm swing.

Jump height in CMJ_{AS} and CMJ_{NAS} from all three methods showed good to excellent intra-session, inter-session, and concurrent reliability. Each method therefore appears reproducible and robust in evaluating CMJ height, with no proportional bias between methods. While proportional bias was generally not observed between FP_{imp} and FP_{time}, it tended to be present or higher in comparisons involving PUSH2 against the other two methods, which aligns with findings from other studies using PUSH2 [20]. Previous studies reported that PUSH2 overestimated CMJ_{NAS} height by 3.8 cm [20] and 7.9 cm [41] compared to the FP_{time}, and 6.0 cm compared to the FP_{imp} [19], which was higher than presently (1.4–1.8 cm from FP_{time} and 3.4–4.0 cm from FP_{imp}). This greater overestimation could be attributed to differences in software versions used across studies. In our study, we used PUSH2 software version 7.7.0. In contrast, Watkins et al. [20] used version 4.5.0, Comyns et al. [41] used version 4.6.2, and McMaster et al. [19] used version 2.0.5. Software updates are typically designed to enhance measurement accuracy; hence, practitioners should be mindful of variations induced by the computation algorithm.

Among limitations to this study, whether the low inter-day reliability of AI_{abs} and AI_{rel} extends to athletes with greater arm contribution must be carefully considered. All participants were men, so whether the present results are applicable to female individuals may need confirmation. Strict control of participants' behavior between sessions was practically not possible, but the high reliability and validity reported suggest that any confounding influence would have been minimal. Lastly, the current findings of PUSH2 might not be applicable to other IMUs likely using a different algorithm, although the hardware technology is essentially the same. Future research may need to clarify how the arm swing impacts the reliability of jump performance measurement across populations of athletes, notably higher-level athletes implementing jumping as a tool to monitor training implementations over longer time periods.

5. Conclusions

All three methods showed high intra-session and inter-session reliability for jump height, supporting their use in athlete monitoring when applied consistently. However, compared to the gold-standard FP_{imp} method, both FP_{time} and PUSH2 overestimated jump height. Because it overestimated jump heights similarly in both CMJ modalities, FP_{time} misestimated the relative effect of arm swing. PUSH2 potentially misestimated both the absolute and relative effects of arm swing, owing to different degrees of overestimation between CMJ_{NAS} and CMJ_{AS}. Also, since arm contribution indices showed lower inter-day reliability, their use in monitoring should be approached with caution.

Author Contributions: Conceptualization, D.Y. and Y.I.; methodology, D.Y.; software, D.Y.; validation, D.Y., Y.I., and F.J.H.; formal analysis, D.Y.; investigation, D.Y. and Y.I.; resources, D.Y.; data curation, D.Y.; writing—original draft preparation, D.Y., Y.I. and F.J.H.; writing—review and editing, D.Y., Y.I. and F.J.H.; visualization, D.Y.; supervision, D.Y.; project administration, D.Y.; funding acquisition, D.Y. All authors have read and agreed to the published version of the manuscript.

Funding: This work was supported by JSPS KAKENHI, Grant Number 19K20003.

Institutional Review Board Statement: This study was conducted in accordance with the Declaration of Helsinki and approved by the Institutional Ethics Committee of the Japan Institute of Sports Sciences (No. 2019-062, approved on 18 February 2020).

Informed Consent Statement: Written informed consent has been obtained from all subjects involved in this study.

Data Availability Statement: The data presented in this study are available upon request from the corresponding author due to institutional policies restricting the public distribution of code and software used in this study.

Conflicts of Interest: The authors declare no conflicts of interest.

Abbreviations

The following abbreviations are used in this manuscript:

AI_{abs}	Absolute arm contribution index (difference in CMJ height between CMJ_{AS} and CMJ_{NAS})
AI_{rel}	Relative arm contribution index (percentage change in CMJ_{AS} height relative to CMJ_{NAS})
CI	Confidence interval
CMJ	Countermovement jump
CMJ_{AS}	Countermovement jump with arm swing
CMJ_{NAS}	Countermovement jump without arm swing
CV	Coefficient of variation
FP_{imp}	Impulse–momentum method using force platforms
FP_{time}	Flight time method using force platforms
GRF_v	Vertical ground reaction force
ICC	Intraclass correlation coefficient
IMU	Inertial measurement unit
LoA	Limits of agreement
PUSH2	PUSH Band 2.0
T_{flight}	Flight time
V_{to}	Vertical velocity at take-off
η_p^2	Partial eta-squared

References

1. Yamashita, D.; Asakura, M.; Ito, Y.; Yamada, S.; Yamada, Y. Physical Characteristics and Performance of Japanese Top-Leve American Football Players. *J. Strength Cond. Res.* **2017**, *31*, 2455–2461. [CrossRef] [PubMed]
2. Teramoto, M.; Cross, C.L.; Rieger, R.H.; Maak, T.G.; Willick, S.E. Predictive Validity of National Basketball Association Draf Combine on Future Performance. *J. Strength Cond. Res.* **2018**, *32*, 396–408. [CrossRef] [PubMed]
3. Edwards, T.; Spiteri, T.; Piggott, B.; Bonhotal, J.; Haff, G.G.; Joyce, C. Monitoring and Managing Fatigue in Basketball. *Sports* **2018** *6*, 19. [CrossRef] [PubMed]
4. Claudino, J.G.; Cronin, J.; Mezencio, B.; McMaster, D.T.; McGuigan, M.; Tricoli, V.; Amadio, A.C.; Serrao, J.C. The Countermovement Jump to Monitor Neuromuscular Status: A Meta-Analysis. *J. Sci. Med. Sport* **2016**, *20*, 397–402. [CrossRef]
5. Sanders, G.J.; Boos, B.; Rhodes, J.; Kollock, R.O.; Peacock, C.A.; Scheadler, C.M. Factors Associated with Minimal Changes ir Countermovement Jump Performance Throughout a Competitive Division I Collegiate Basketball Season. *J. Sports Sci.* **2019**, *37* 2236–2242. [CrossRef]
6. Heishman, A.D.; Daub, B.D.; Miller, R.M.; Freitas, E.D.S.; Frantz, B.A.; Bemben, M.G. Countermovement Jump Reliability Performed with and without an Arm Swing in Ncaa Division 1 Intercollegiate Basketball Players. *J. Strength Cond. Res.* **2020**, *34* 546–558. [CrossRef]
7. Borràs, X.; Balius, X.; Drobnic, F.; Galilea, P. Vertical Jump Assessment on Volleyball: A Follow-Up of Three Seasons of a High-Level Volleyball Team. *J. Strength Cond. Res.* **2011**, *25*, 1686–1694. [CrossRef]
8. Gonçalves, C.A.; Lopes, T.J.D.; Nunes, C.; Marinho, D.A.; Neiva, H.P. Neuromuscular Jumping Performance and Upper-Body Horizontal Power of Volleyball Players. *J. Strength Cond. Res.* **2021**, *35*, 2236–2241. [CrossRef]
9. Wen, N.; Dalbo, V.J.; Burgos, B.; Pyne, D.B.; Scanlan, A.T. Power Testing in Basketball: Current Practice and Future Recommendations. *J. Strength Cond. Res.* **2018**, *32*, 2677–2691. [CrossRef]
10. Chiu, L.Z.F.; Bryanton, M.A.; Moolyk, A.N. Proximal-to-Distal Sequencing in Vertical Jumping With and Without Arm Swing. *J. Strength Cond. Res.* **2014**, *28*, 1195–1202. [CrossRef]
11. Vaverka, F.; Jandačka, D.; Zahradník, D.; Uchytil, J.; Farana, R.; Supej, M.; Vodičar, J. Effect of an Arm Swing on Countermovement Vertical Jump Performance in Elite Volleyball Players: FINAL. *J. Hum. Kin.* **2016**, *53*, 41–50. [CrossRef] [PubMed]
12. Gerodimos, V.; Zafeiridis, A.; Perkos, S.; Dipla, K.; Manou, V.; Kellis, S. The Contribution of Stretch-Shortening Cycle and Arm-Swing to Vertical Jumping Performance in Children, Adolescents, and Adult Basketball Players. *Pediatr. Exerc. Sci.* **2008**, *20* 379–389. [CrossRef]
13. Lake, J.P.; Mundy, P.; Comfort, P.; McMahon, J.J.; Suchomel, T.J.; Carden, P. Concurrent Validity of a Portable Force Plate Using Vertical Jump Force-Time Characteristics. *J. Appl. Biomech.* **2018**, *34*, 410–413. [CrossRef]
14. Yamashita, D.; Murata, M.; Inaba, Y. Effect of Landing Posture on Jump Height Calculated from Flight Time. *Appl. Sci.* **2020**, *10* 776. [CrossRef]
15. Rago, V.; Brito, J.; Figueiredo, P.; Carvalho, T.; Fernandes, T.; Fonseca, P.; Rebelo, A. Countermovement Jump Analysis Using Different Portable Devices: Implications for Field Testing. *Sports* **2018**, *6*, 91. [CrossRef]
16. Aragón, L.F. Evaluation of Four Vertical Jump Tests: Methodology, Reliability, Validity, and Accuracy. *Meas. Phys. Edu. Exer. Sci* **2000**, *4*, 215–228. [CrossRef]
17. Willy, R.W. Innovations and pitfalls in the use of wearable devices in the prevention and rehabilitation of running related injuries *Phys. Ther. Sport* **2018**, *29*, 26–33. [CrossRef]
18. Windt, J.; MacDonald, K.; Taylor, D.; Zumbo, B.D.; Sporer, B.C.; Martin, D.T. "To Tech or Not to Tech?" A Critical Decision-Making Framework for Implementing Technology in Sport. *J. Athl. Train.* **2020**, *55*, 902–910. [CrossRef]
19. McMaster, D.T.; Tavares, F.; O'Donnell, S.; Driller, M. Validity of Vertical Jump Measurement Systems. *Meas. Phys. Edu. Exer. Sci* **2021**, *25*, 95–100. [CrossRef]
20. Watkins, C.M.; Maunder, E.; Tillaar, R.V.D.; Oranchuk, D.J. Concurrent Validity and Reliability of Three Ultra-Portable Vertical Jump Assessment Technologies. *Sensors* **2020**, *20*, 7240. [CrossRef]
21. Montalvo, S.; Gonzalez, M.P.; Dietze-Hermosa, M.S.; Eggleston, J.D.; Dorgo, S. Common Vertical Jump and Reactive Strength Index Measuring Devices: A Validity and Reliability Analysis. *J. Strength Cond. Res.* **2021**, *35*, 1234–1243. [CrossRef] [PubMed]
22. Cohen, J. *Statistical Power Analysis for the Behavioral Sciences*, 2nd ed.; Routledge: Abingdon, UK, 1988. [CrossRef]
23. Iglesias-Caamaño, M.; Álvarez-Yates, T.; Carballo-López, J.; Cuba-Dorado, A.; García-García, O. Interday Reliability of a Testing Battery to Assess Lateral Symmetry and Performance in Well-Trained Volleyball Players. *J. Strength Cond. Res.* **2022**, *36*, 895–901 [CrossRef] [PubMed]
24. Cormack, S.J.; Newton, R.U.; McGuigan, M.R.; Doyle, T.L.A. Reliability of Measures Obtained During Single and Repeated Countermovement Jumps. *Int. J. Sports Physiol. Perform.* **2008**, *3*, 131–144. [CrossRef] [PubMed]

5. Dobbin, N.; Hunwicks, R.; Highton, J.; Twist, C. A Reliable Testing Battery for Assessing Physical Qualities of Elite Academy Rugby League Players. *J. Strength Cond. Res.* **2018**, *32*, 3232–3238. [CrossRef]
6. Heishman, A.D.; Curtis, M.A.; Saliba, E.N.; Hornett, R.J.; Malin, S.K.; Weltman, A.L. Comparing Performance During Morning vs. Afternoon Training Sessions in Intercollegiate Basketball Players. *J. Strength Cond. Res.* **2017**, *31*, 1557–1562. [CrossRef]
7. Street, G.; McMillan, S.; Board, W.; Rasmussen, M.; Heneghan, J.M. Sources of Error in Determining Countermovement Jump Height with the Impulse Method. *J. Appl. Biomech.* **2001**, *17*, 43–54. [CrossRef]
8. Sado, N.; Yoshioka, S.; Fukashiro, S. Free-leg side elevation of pelvis in single-leg jump is a substantial advantage over double-leg jump for jumping height generation. *J. Biomech.* **2020**, *104*, 109751. [CrossRef]
9. Bland, J.M.; Altman, D.G. Statistical Methods for Assessing Agreement Between Two Methods of Clinical Measurement. *Lancet* **1986**, *327*, 307–310. [CrossRef]
0. Atkinson, G.; Nevill, A.M. Statistical Methods for Assessing Measurement Error (Reliability) in Variables Relevant to Sports Medicine. *Sports Med.* **1998**, *26*, 217–238. [CrossRef]
1. Koo, T.K.; Li, M.Y. A Guideline of Selecting and Reporting Intraclass Correlation Coefficients for Reliability Research. *J. Chiropr. Med.* **2016**, *15*, 155–163. [CrossRef]
2. Bland, J.M.; Altman, D.G. Applying the Right Statistics: Analyses of Measurement Studies. *Ultrasound Obstet. Gynecol.* **2003**, *22*, 85–93. [CrossRef] [PubMed]
3. Bland, J.M.; Altman, D.G. Agreement between methods of measurement with multiple observations per individual. *J. Biopharm. Stat.* **2007**, *17*, 571–582. [CrossRef] [PubMed]
4. García-Ramos, A.; Pérez-Castilla, A.; Martín, F. Reliability and Concurrent Validity of the Velowin Optoelectronic System to Measure Movement Velocity During the Free-Weight Back Squat. *Int. J. Sports Sci. Coach.* **2018**, *13*, 737–742. [CrossRef]
5. Hopkins, W.G. A Scale of Magnitudes for Effect Statistics: A New View of Statistics. *Sportscience* **2002**, *502*, 411.
6. Lake, J.P.; Augustus, S.; Austin, K.; Mundy, P.; McMahon, J.J.; Comfort, P.; Haff, G.G. The Validity of the Push Band 2.0 During Vertical Jump Performance. *Sports* **2018**, *6*, 140. [CrossRef]
7. Lajimi, S.A.M.; McPhee, J. A comprehensive filter to reduce drift from Euler angles, velocity, and position using an IMU. In Proceedings of the 2017 IEEE 30th Canadian Conference on Electrical and Computer Engineering (CCECE), Windsor, ON, Canada, 30 April–3 May 2017; pp. 1–6.
8. Hara, M.; Shibayama, A.; Takeshita, D.; Fukashiro, S. The Effect of Arm Swing on Lower Extremities in Vertical Jumping. *J. Biomech.* **2006**, *39*, 2503–2511. [CrossRef]
9. Feltner, M.E.; Fraschetti, D.J.; Crisp, R.J. Upper Extremity Augmentation of Lower Extremity Kinetics during Countermovement Vertical Jumps. *J. Sports Sci.* **1999**, *17*, 449–466. [CrossRef] [PubMed]
0. Bishop, C.; Shrier, I.; Jordan, M. Ratio Data: Understanding Pitfalls and Knowing When to Standardise. *Symmetry* **2023**, *15*, 318. [CrossRef]
1. Comyns, T.M.; Murphy, J.; O'Leary, D. Reliability, Usefulness, and Validity of Field-Based Vertical Jump Measuring Devices. *J. Strength Cond. Res.* **2023**, *37*, 1594–1599. [CrossRef]

biomechanics

MDPI

Article

Application of Synchronized Inertial Measurement Units and Contact Grids in Running Technique Analysis: Reliability and Sensitivity Study †

Đorđe Brašanac, Marko Kapeleti *, Igor Zlatović, Miloš Ubović and Vladimir Mrdaković

Faculty of Sport and Physical Education, University of Belgrade, Blagoja Parovića 156, 11030 Belgrade, Serbia; brasanacps@gmail.com (Đ.B.); igor.zlatovic@fsfv.bg.ac.rs (I.Z.); ubovicmilos@yahoo.com (M.U.); vladimir.mrdakovic@fsfv.bg.ac.rs (V.M.)

* Correspondence: marko.kapeleti@fsfv.bg.ac.rs

† This paper is an extended version of our conference abstract: Brašanac, Đ.; Kapeleti, M.; Zlatović, I.; Ubović, M.; Mrdaković, V. (2025). Kinematic and kinetic analysis of the Groucho running technique using inertial measurement units and contact detection devices. In Book of Abstracts from 22nd International Scientific Conference on Physical Activity and Health—Pledge for Life (p. 67). Faculty of Sport and Physical Education, Belgrade, Serbia, 6–7 December 2024; Dabović, M., Bubanja, I., Miletić, V., Eds. The link of the conference: https://www.fsfvconference.rs/ (accessed on 18 September 2025).

Abstract: Background: Previous research has identified center of mass vertical oscillation and leg stiffness as the most common variables differentiating Natural and Groucho running techniques. The aim was to assess the inter-session reliability and inter-technique sensitivity of synchronized inertial measurement units and contact grids in quantifying kinematic and kinetic differences between Natural and Groucho running techniques. **Methods:** Eleven physically active and healthy males ran at a speed 50% higher than transition speed. Two sessions for Natural and two for Groucho running were performed, each lasting 1 min. **Results:** Most variables exhibited a similar inter-session reliability across running techniques, except contact time and center of mass vertical displacement, ranging from moderate to good (ICC = 0.538–0.897). A statistically significant difference between running techniques was found for all variables ($p < 0.05$), except for contact time and center of mass vertical oscillation ($p > 0.05$), likely due to inconsistency in reliability depending on the running technique, which may have covered the underlying differences. **Conclusions:** We can conclude that the combination of synchronized inertial measurement units and contact grids showed potentially acceptable reliability and sufficient sensitivity to recognize and differentiate between Natural and Groucho running techniques. The results may contribute to a broader understanding of the differences between these two running techniques and encourage the increased use of these devices within therapeutic, recreational, and sports running contexts.

Keywords: natural running; Groucho running; center of mass vertical oscillation; leg stiffness; kinematics; kinetics

1. Introduction

Inertial measurement units (IMUs) are multicomponent sensor systems designed to measure acceleration, angular velocity, and the orientation of a physical body in space. These systems consist of three key sensors—an accelerometer, a gyroscope, and a magnetometer—that collectively enable the estimation of kinematic parameters. The accelerometer measures linear acceleration along three mutually perpendicular axes by detecting the displacement of an inertial mass relative to a reference frame. The gyroscope measures the angular velocity around three axes, whereas the magnetometer detects the strength and direction of the Earth's magnetic field, allowing for the determination of the device's absolute orientation in space. The reliability of IMUs has been confirmed in numerous studies examining different types of movements, such as walking, jumping [1], running [2] and various weightlifting exercises—Olympic lifts, squats, etc. [1,3]. By enabling detailed tracking of movement and contact phases, the use of IMUs helps identify movement patterns and optimize techniques to improve performance and reduce the risk of injury [4–7].

Contact grids (CGs) consist of two units, which together create a "carpet" of infrared light a few millimeters above the floor. When the light beams are interrupted, the time is recorded, thus allowing detection of contact and flight times. These devices have a wide range of applications in biomechanical analysis, such as running and jumping [8–10]. The IMU sensor can be combined with other measuring devices—such as, for example, CGs—in order to achieve more precise and comprehensive motion analysis. If synchronized within the common software, they can offer broader spectrum of already predefined and precalculated variables, facilitating the comprehensiveness and time efficiency of analysis when using these devices in practical settings. Thus, the combination of IMUs and CGs could further expand the scope of biomechanical movement analysis, offering broader applications and deeper insights.

Running is a fundamental form of human locomotion that is often analyzed from a biomechanical perspective to better understand its basic principles. Research in this field has focused on the forces acting on the body, movement patterns, and factors influencing performance [11–14]. Vertical oscillation of the body's center of mass is one of the most commonly used reference points in the biomechanical analysis of running [15–17]. Another important running factor is the stiffness of the locomotor system, defined as the extent to which an object resists deformation in response to an applied force [18], which reflects the ability to accumulate, store, and release energy from elastic deformation [19]. All of this could be analyzed with a combination of IMUs and CGs by utilizing their technological ability to estimate displacements and ground reaction forces during movement on the basis of acceleration and contact detection. Since acceleration is the second derivative of displacement, integrating the acceleration signal twice (with proper filtering) can yield estimates of velocity and displacement. On the other hand, contact time, measured using CGs, serves as a basis for assessing oscillatory or spring-like behavior of the body, particularly during running.

An optimal level of leg stiffness can enhance running performance [20] and reduce injury occurrence [21]. Runners frequently use various strategies to optimize stiffness, including modifying surfaces, footwear, and running techniques [22,23], such as the Groucho running technique [20]. This technique is characterized by reduced vertical oscillation of the center of mass, lower leg stiffness, and increased hip and knee flexion [24], thus reducing loading on the musculoskeletal system and increasing muscle activation and oxygen consumption [25,26], potentially helping to prevent injuries while enhancing the effectiveness of training within a specific timeframe.

Research on the biomechanical differences between the Natural and Groucho running techniques has shown that the Groucho technique is associated with a longer contact time, shorter flight phase [27,28], shorter stride length [24], higher step frequency [28], reduced vertical oscillation of the center of mass [24], and lower leg stiffness [24]. Most previous studies have analyzed and compared these two running techniques via motion capture systems [27] or a combination of kinematic analysis and force platforms [29]. Despite the widespread use of IMU and CG devices in the biomechanical analysis of running techniques, the analysis and comparison of Natural and Groucho running techniques has not yet been examined using these devices.

The aim of this study is to assess the inter-session reliability and inter-technique sensitivity of IMUs and CGs in quantifying previously established differences between Natural and Groucho running techniques. Given the technological characteristics and physics principles underlying these devices, it is hypothesized that the measurements will demonstrate an acceptable level of reliability and that they will be sufficiently sensitive to detect differences in kinematic and kinetic characteristics between these two techniques. The results may contribute to a broader understanding of the differences between these two running techniques and encourage the increased use of these devices within therapeutic, recreational, and sports running contexts by facilitating more comprehensive and time efficient analysis when used in practical settings.

2. Materials and Methods

2.1. Sample of Subjects

The study included 11 physically active and healthy male participants without recent injuries to the locomotor system and previously inexperienced in Groucho running. The participants' age was 23.1 $\pm$ 2.5 years, body height was 182.7 $\pm$ 5.8 cm, body mass was 83.1 $\pm$ 7.4 kg, and body mass index was 24.9 $\pm$ 1.6 kg/m^2. Each subject signed a written informed consent form confirming their voluntary participation in the experiment. Achieved statistical power for this sample size was calculated using G*Power software v3.1.9.7 (Heinrich Heine University, Düsseldorf, Germany). Analysis for two-tailed dependent (paired) T-test showed that 11 participants produced moderate 67% statistical power in detecting large effect size f (0.4), at an alpha level of 0.05.

2.2. Experimental Protocol

The protocol consisted of basic anthropometric and morphological measurements (body height, body mass, and dominant leg length), followed by a ten-minute warm-up through cyclic exercises, stretching, and muscle activation drills. The next step involved determining the transition speed from walking to running, followed by familiarization with the Groucho running technique, and finally, the main part of the experimental protocol—variable sampling during Natural and Groucho running techniques. The subjects were instructed not to consume food or liquids for at least one hour before the testing session and not to engage in any intense physical activity for at least 24 h before the experiment. The experiments reported in the article were undertaken in compliance with the relevant laws and institutional guidelines. The experimental protocol was planned and conducted in accordance with the Declaration of Helsinki and was approved by the Ethics Committee of the Faculty of Sport and Physical Education, University of Belgrade (Date: 9 April 2025; Number: 02-506/25-2).

Body height was measured via a digital altimeter (BSM170, Arab Engineers). The measurement procedure was conducted following the manufacturer's recommendations. Body mass index was calculated as the ratio between body mass [kg] and body height [m] squared. Leg length measurement was required for the calculation of leg stiffness via software (see the Variables section).

Running speed was standardized across participants based on transition speed from walking to running in order to minimize the differences in Natural running technique between the subjects, as transition speed accounts for variations in anthropometric characteristics and motor abilities important for running at the particular speed [30]. The transition speed determination protocol involved testing under natural walking conditions on a treadmill while continuously holding the handles. The participants were instructed that the goal was not to determine their maximum walking speed but rather the speed at which walking or running felt more natural and easier. The treadmill display showing speed was hidden, and participants were free to choose the more suitable movement pattern for a given speed. The initial walking speed was 5 km/h, and the speed was gradually increased every 30 s by either 0.4 km/h or 0.2 km/h depending on the proximity to the transition speed. The objective indicator of transition speed was the presence of a flight phase, whereas the subjective indicator was the participant's feedback. After the initial transition speed determination, a verification test was performed at the given speed following a 30 s break. In this sample of subjects, the transition speed was 7.1 ± 0.4 km/h.

Familiarization with the Groucho running technique was conducted through verbal instructions, demonstrations, participant practice, and corrections made by the examiner. The criteria for proper Groucho running technique included visibly reduced vertical oscillation of the center of mass, continuous maintenance of the center of mass at a constant level, and increased flexion at the hip and knee joints of both legs. The exact verbal instructions were as follows: "Flex your knees, lower your body height while running, keep the height at a constant level and reduce the impact of the foot". Familiarization lasted around 5–10 min, depending on the subject learning efficiency, until the subject could perform proper Groucho technique continuously and independently. Familiarization for Natural running technique was unnecessary because it did not involve any verbal cues in order to simulate real natural conditions for an individual.

The participants ran at a supratransitional speed—their individual transition speed increased by 50%, to reduce the variability of relatively slow running on transition speed, as it is known that increases in running speed reduces the coordination variability during stance phase [31]. For this sample of subjects, the running speed was 10.6 ± 0.5 km/h. Two running sessions were performed per technique in the following order: (1) Natural running and (2) Groucho running, each lasting 1 min, with a 2 min break between techniques and a 10 min break between sessions. Techniques were not randomized within a session across participants because Natural running was performed without verbal instructions, so it did not affect performance during Groucho running as there is no learning and transfer effect, as well as fatigue occurrence due to controlled running intensity and the duration of resting period. Technique control during the experiment was conducted through visual observation, with corrections provided via previously explained verbal instructions from the examiner.

2.3. Testing Devices

Two complementary measuring devices (IMU and CG) made by the Norwegian company MuscleLab (Figure 1) were used. The IMU consisted of an accelerometer (16G, accuracy ±1%), a gyroscope (2000 dps, accuracy ±1%), and a magnetometer (1300/2500 μT, accuracy ±5%). The sampling frequency of the IMU was 200 Hz. The resolution of the CG was < 2 ms. Synchronization was achieved via a data synchronization unit (accuracy ±1 ms) with a built-in radio host device (2.4 GHz), whereas the software enabled data acquisition and processing (v10.241.11.4.5359, x64 bit). The IMU sensor was placed at the L5–S1 level to approximate the center of mass location, whereas the CGs were placed in line with the treadmill and separated 3 m apart.

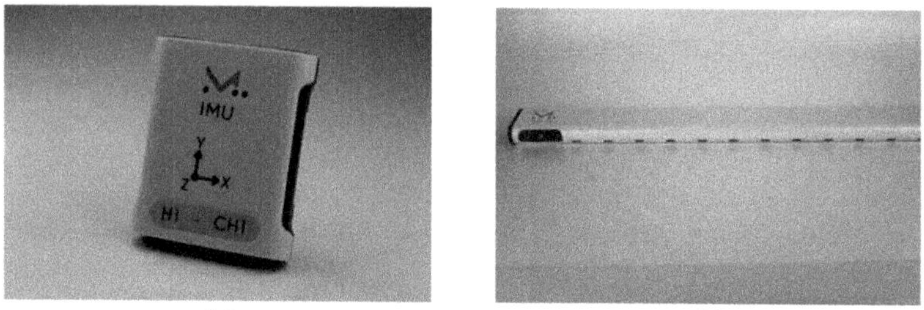

(**a**) (**b**)

Figure 1. Inertial measurement unit (**a**) and contact grid (**b**) made by the Norwegian company MuscleLab https://www.musclelabsystem.com/products/ (accessed on 1 October 2025).

2.4. Variables

The following variables were acquired:

(1) Contact time—CT [ms]
(2) Flight time—FT [ms]
(3) Step length—SL [cm]
(4) Step frequency—SF [step/s]
(5) Center of mass takeoff angle—COM_{angle} [°]
(6) Center of mass vertical oscillation—O_{vert} [cm]
(7) Leg stiffness—K_{leg} [kN/m]

CT, FT, SL and SF are the most common kinematic variables used for running analysis [32,33]. The COM_{angle} refers to the angle between the horizontal axis and the resultant velocity vector of the COM at the moment the subject leaves the ground. We utilized this variable in order to describe the differences between techniques in a more comprehensive manner, given that different takeoff angle under the same impulse influence the achieved height of the object in flight, and consequently O_{vert}. Software computed K_{leg} on the basis of the spring-mass model, utilizing leg length and joint kinematics. Specifically, leg compression (ΔL) was estimated by calculating changes in effective leg length during the stance phase on the basis of initial leg length, running velocity and contact time [34] which were all derived from predefined anthropometric data and velocity, as well as IMU-measured body motion and CG detection. The peak vertical force (F_{max}) was approximated via the vertical displacement recorded by the IMU and determined via double integration of the vertical acceleration over time [35]. K_{leg} was then computed as the ratio of F_{max} to ΔL [34].

2.5. Statistical Analysis

Each variable was represented by its mean value and standard deviation. Data were subjected to normality distribution analysis via the Shapiro–Wilk test (for both sessions and techniques). Reliability analysis was conducted by comparing both sessions within technique via the average measures intraclass correlation coefficient (ICC) with a two-way mixed-effects model and absolute agreement, presented with a 95% confidence interval (CI) and standard error of measurement (SEM). The following criteria were used for ICC classification: poor < 0.50, moderate 0.50–0.74, good 0.75–0.89, and excellent $\geq$ 0.90 [36]. Systematic errors within technique were examined through repeated-measures analysis of variance. Sensitivity analysis was conducted by comparing the second sessions of both techniques via dependent T-test. Effect size (ES) for the differences between running techniques was calculated via Cohen's D as the difference between the mean values divided by the pooled standard deviation. The effect size was considered small if it was $\leq$0.49, medium if it was between 0.50 and 0.79, and large if it was $\geq$0.80 [37]. Statistical data processing was performed via IBM SPSS Statistics v20.0.0 (IBM Corp., Armonk, NY, USA). The level of statistical significance was set at $\leq$0.05.

3. Results

The Shapiro–Wilk test confirmed a normal distribution of data for all variables in both sessions and for both techniques in this sample of subjects ($p > 0.05$). The results of the repeated-measures analysis of variance within technique did not show significance, indicating that no systematic error was detected ($p > 0.05$). Figure 2 shows the representative plots of O_{vert} and K_{leg} during 10 s across running techniques. Figures 3 and 4 show the descriptive statistics for all the variables across running techniques. Tables 1 and 2 present the results of the inter-session reliability analysis for the Natural and Groucho running techniques, respectively. Table 3 presents the results of the inter-technique sensitivity analysis.

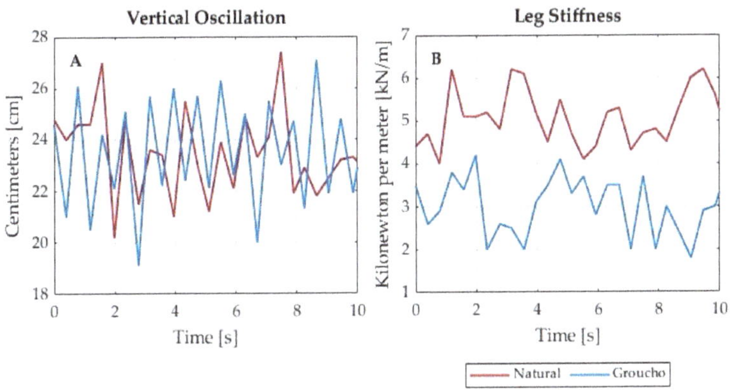

Figure 2. Representative plot for (**A**) vertical oscillation and (**B**) leg stiffness during Natural and Groucho running.

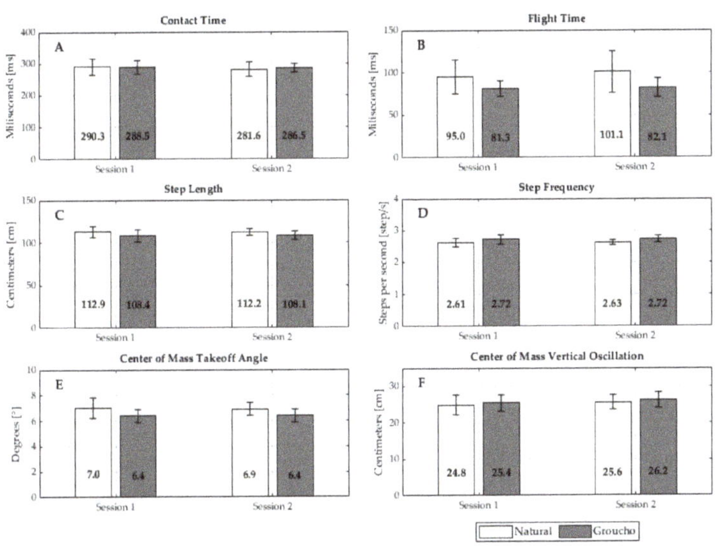

Figure 3. Means ± standard deviations for all kinematic variables during Natural and Groucho running—(**A**) contact time, (**B**) flight time, (**C**) step length, (**D**) step frequency, (**E**) center of mass takeoff angle and (**F**) center of mass vertical oscillation.

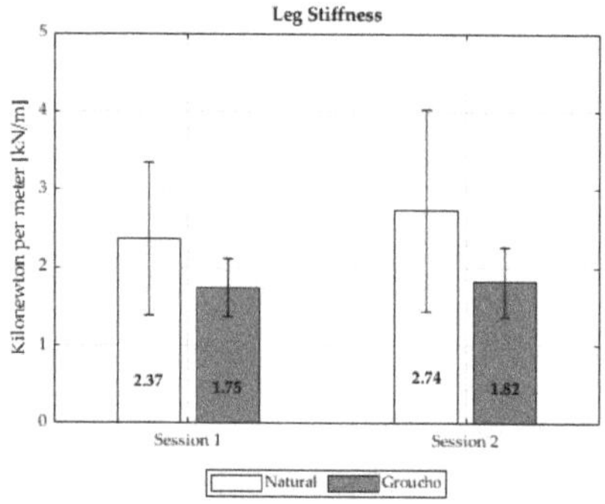

Figure 4. Means ± standard deviations for leg stiffness during Natural and Groucho running.

Table 1. Results of the ICC and SEM analysis for the Natural running technique.

Variables	ICC	95% CI	SEM	p
CT	0.575	−0.487–0.884	17.9	0.095
FT	0.690	−0.114–0.916	17.5	0.040 *
SL	0.897	0.626–0.972	2.5	0.001 *
SF	0.845	0.433–0.958	0.07	0.004 *
COM_{angle}	0.884	0.566–0.969	0.3	0.001 *
O_{vert}	0.684	−0.084–0.913	1.3	0.039 *
K_{leg}	0.720	0.030–0.923	0.88	0.027 *

CT—contact time; FT—flight time; SL—step length; SF—step frequency; COM_{angle}—center of mass takeoff angle; O_{vert}—COM vertical oscillation; K_{leg}—leg stiffness. * Statistical significance ($p \leq 0.05$).

Table 2. Results of the ICC and SEM analysis for the Groucho running technique.

Variables	ICC	95% CI	SEM	p
CT	0.843	0.410–0.958	12.9	0.005 *
FT	0.622	−0.546–0.901	17.4	0.082
SL	0.883	0.555–0.969	1.8	0.002 *
SF	0.800	0.216–0.947	0.04	0.012 *
COM_{angle}	0.822	0.340–0.952	0.4	0.007 *
O_{vert}	0.868	0.516–0.964	0.9	0.001 *
K_{leg}	0.538	−0.858–0.878	0.88	0.131

CT—contact time; FT—flight time; SL—step length; SF—step frequency; COM_{angle}—center of mass takeoff angle; O_{vert}—COM vertical oscillation; K_{leg}—leg stiffness. * Statistical significance ($p \leq 0.05$).

Table 3. Results of dependent T-test and effect size analysis between running techniques.

Variables	t	p	Cohen's D
CT	−0.802	0.441	0.26
FT	+2.663	0.024 *	1.00
SL	+5.747	0.000 *	0.90
SF	−5.078	0.000 *	0.94
COM_{angle}	+4.627	0.001 *	1.00
O_{vert}	−1.464	0.174	0.30
K_{leg}	+2.469	0.033 *	0.95

CT—contact time; FT—flight time; SL—step length; SF—step frequency; COM_{angle}—center of mass takeoff angle; O_{vert}—COM vertical oscillation; K_{leg}—leg stiffness. * Statistical significance ($p \leq 0.05$).

For almost all the variables, the ICC was statistically significant, except for CT, which tended toward statistical significance. The ICCs ranged from moderate to good for all the variables (0.575–0.897). In 3 out of 7 cases, inter-session reliability was good. Moderate reliability was observed for CT, FT, O_{vert}, and K_{leg}, whereas good reliability was observed for SL, SF, and COM_{angle}.

For almost all the variables, the ICCs were statistically significant. K_{leg} was not statistically significant, whereas FT tended toward statistical significance. The ICCs ranged from moderate to good for all the variables (0.538–0.883). In 5 out of 7 cases, inter-session reliability was good. Moderate reliability was observed for FT and K_{leg}, whereas good reliability was observed for CT, SL, SF, COM_{angle}, and O_{vert}.

A statistically significant difference between running techniques was found for all the variables, except for CT and O_{vert}. Figures 3 and 4 show that FT, SL, COM_{angle}, and K_{leg} are lower, whereas SF is higher in the Groucho running technique.

4. Discussion

The aim of this study was to examine the inter-session reliability and inter-technique sensitivity of IMUs and CGs in detecting differences in kinematic and kinetic characteristics between Natural and Groucho running techniques. The devices demonstrated, in most cases, nearly good-to-good reliability in measurements and are sufficiently sensitive to detect differences in most of the monitored variables, which could be considered adequate in the measurements of biological systems. The significance of this study lies in providing new insights into the application of synchronized IMUs and CGs in both scientific research and practical fields related to running, as this configuration may enhance capability and efficiency of running analysis.

Most variables exhibited a statistically significant ICC, with inter-session reliability ranging from moderate to good (ICC = 0.538–0.897) (Tables 1 and 2). Specifically, moderate reliability in both techniques is demonstrated for FT and K_{leg}, good reliability is demonstrated in both techniques for SL, SF and COM_{angle}, and the reliability of CT and O_{vert} differs from moderate to good depending on the running technique. No variables reached poor level of reliability. As previously mentioned, the adequate reliability of these devices has been confirmed in numerous studies examining different movement patterns, such as walking and jumping [1], running [2], and various weightlifting movements, including Olympic lifts and squats [1,3].

Previous research on this topic has identified O_{vert} and K_{leg} as the most common variables differentiating these two running techniques [24,38]. The present results reveal a statistically significant difference between the running techniques for most analyzed variables, except for CT and O_{vert} (Table 3), probably because of inconsistency in reliability depending on the running technique, and relatively lower reliability in the Natural compared to Groucho running technique (Tables 1 and 2). ESs of all variables are in accordance with the statistical significance, that is, variables that showed significant differences between techniques had large ES (Table 3). The lower FT, SL, COM_{angle}, and K_{leg} values, along with an increased SF in the Groucho technique, suggest that this technique involves shorter, more frequent, and "softer" steps, which aligns with previous research findings [24,27,28].

Previous studies have reported that the Groucho running technique results in longer CT and lower O_{vert} values [24,27,28,38]. A partial explanation for the lack of observed differences in this study is the inconsistency in the reliability of CT and O_{vert} across running techniques (predominantly in Natural), which may have covered the underlying differences. Regarding O_{vert}, this finding is not in line with previous studies that confirmed the reliability of this variable [39], while CT has the tendency to show relatively lower validity and reliability via IMUs [40,41]. Additionally, it is hypothesized that the reduced K_{leg} in the Groucho technique alters vertical oscillation—being more pronounced in the eccentric phase and less pronounced in the concentric phase than in the Natural running technique—which may partially explain this unexpected result. In support of this hypothesis, the lower COM_{angle} observed in the Groucho technique likely reduces the height of the COM trajectory, leading to less upward vertical oscillation after takeoff. Furthermore, it is unlikely that this unexpected result stems from methodological issues, as participants had sufficient familiarization that enabled them to include every technical request in the Groucho running described in the exact verbal instructions, data from the second session were used for sensitivity analysis, the IMU was positioned on an appropriate location and equipment was well synchronized. The authors of this study do not necessarily imply that CT and O_{vert} are not suitable for distinguishing the differences between Natural and Groucho running techniques and running technique analysis in general, but that these variables should be re-examined on reliability in different research settings along with joint kinematic analysis to describe the movement more comprehensively.

Since the only way to validate the accuracy of these devices is to directly compare the differences in biomechanical variables between these running techniques, with findings from previous studies that employed more direct measurement methods and given the previously stated explanations for CT and O_{vert}, we can conclude that the combination of synchronized IMUs and CGs showed potentially acceptable reliability and sufficient sensitivity to recognize and differentiate between Natural and Groucho running techniques. This study contributes to a better understanding of the capabilities and potential applications of synchronized IMUs and CGs for biomechanical running analysis, as this technological configuration has the potential to analyze efficiently more variables than using, for example, only CGs for running technique analysis. The primary uncertainty pertains to the CT and O_{vert} variables, which requires further investigation. One of the study limitations is the relatively smaller sample size, leading to moderate statistical power (67%) and lower chance of detecting true differences (risk of type II error), due to which we cannot confidently conclude if the differences in CT and O_{vert} between running techniques ultimately exist or not. Reflecting on the limitations of this study, future research should include a larger sample size and a population of professional runners already familiar with the Groucho running technique, in order to examine the generalizability of these findings. Also, it is suggested to explore potential differences among different device manufacturers and across different movements, in order to confirm the accuracy of these findings. Moreover, it is crucial to validate their use in this context by comparing them with the gold standard—motion capture systems.

5. Conclusions

All variables monitored through synchronized IMU and CG exhibited moderate to good inter-session reliability, while most of them were sensitive enough to differentiate between Natural and Groucho running techniques. According to these results, we can conclude that synchronized IMUs and CGs showed potentially acceptable reliability and sufficient sensitivity to recognize and differentiate between Natural and Groucho running techniques, which could be considered adequate in the measurement of biological systems. This study contributed to a better understanding of the potential of utilizing synchronized IMUs and CGs, thereby encouraging their broader application in running-related research and practice. Future studies should re-evaluate the measurement properties of synchronized IMUs and CGs and compare it with gold standard (motion capture system), while utilizing different device manufacturers on a larger sample size of professional runners already familiar with the Groucho running technique.

Author Contributions: Conceptualization, M.K. and V.M. Data curation, Đ.B. and M.K. Formal analysis, Đ.B., M.K. and I.Z. Investigation, Đ.B., M.K., I.Z. and M.U. Methodology, M.K., I.Z. and V.M. Project administration, Đ.B. and M.U. Resources, V.M. Supervision, V.M. Validation, Đ.B., M.K., I.Z. and M.U. Visualization, Đ.B. and M.U. Writing—original draft, Đ.B. and M.K. Writing—review and editing, I.Z., M.U. and V.M. All authors have read and agreed to the published version of the manuscript.

Funding: No financial or material support of any kind was received for the work described in this article.

Institutional Review Board Statement: The experiments reported in the article were undertaken in compliance with the relevant laws and institutional guidelines. The experimental protocol was planned and conducted in accordance with the Declaration of Helsinki and was approved by the Ethics Committee of the Faculty of Sport and Physical Education, University of Belgrade (Date: 9 April 2025; Number: 02-506/25-2).

Informed Consent Statement: Each subject signed a written informed consent form confirming their voluntary participation in the experiment.

Data Availability Statement: The raw data supporting the conclusions of this article will be made available by the authors on request.

Conflicts of Interest: The authors declare no potential or actual conflict of interest that could inappropriately influence (bias) their work.

References

1. Al-Amri, M.; Nicholas, K.; Button, K.; Sparkes, V.; Sheeran, L.; Davies, J.L. Inertial measurement units for clinical movement analysis: Reliability and concurrent validity. *Sensors* **2018**, *18*, 719. [CrossRef]
2. Zeng, Z.; Liu, Y.; Hu, X.; Tang, M.; Wang, L. Validity and reliability of inertial measurement units on lower extremity kinematics during running: A systematic review and meta-analysis. *Sports Med. Open* **2022**, *8*, 86. [CrossRef]
3. Clemente, F.M.; Akyildiz, Z.; Pino-Ortega, J.; Rico-González, M. Validity and reliability of the inertial measurement unit for barbell velocity assessments: A systematic review. *Sensors* **2021**, *21*, 2511. [CrossRef]
4. Reenalda, J. Current Developments and Future Directions in Using IMUs for Injury Prevention in Running. In Proceedings of the 2023 IEEE SENSORS, Vienna, Austria, 29 October—1 November 2023; IEEE: Piscataway, NJ, USA, 2023; pp. 1–2. [CrossRef]
5. Neal, B.S.; Bramah, C.; McCarthy-Ryan, M.F.; Moore, I.S.; Napier, C.; Paquette, M.R.; Gruber, A.H. Using wearable technology data to explain recreational running injury: A prospective longitudinal feasibility study. *Phys. Ther. Sport* **2024**, *65*, 130–136. [CrossRef]
6. McDevitt, S.; Hernandez, H.; Hicks, J.; Lowell, R.; Bentahaikt, H.; Burch, R.; Ball, J.; Chander, H.; Freeman, C.; Taylor, C.; et al. Wearables for biomechanical performance optimization and risk assessment in industrial and sports applications. *Bioengineering* **2022**, *9*, 33. [CrossRef]
7. Xu, D.; Zhou, H.; Jie, T.; Zhou, Z.; Yuan, Y.; Jemni, M.; Quan, W.; Gao, Z.; Xiang, L.; Gusztav, F.; et al. Data-driven deep learning for predicting ligament fatigue failure risk mechanisms. *Int. J. Mech. Sci.* **2025**, *301*, 110519. [CrossRef]
8. Vecbērza, L.; Šmite, Z.; Plakane, L.; Ābeļkalns, I. The impact of ankle plantar-flexor muscle strength on sprint acceleration in floorball players. *Int. J. Sports Physiol. Perform.* **2025**, *20*, 393–398. [CrossRef] [PubMed]
9. Falch, H.N.; Kristiansen, E.L.; Haugen, M.E.; van den Tillaar, R. Association of performance in strength and plyometric tests with change of direction performance in young female team-sport athletes. *J. Funct. Morphol. Kinesiol.* **2021**, *6*, 83. [CrossRef]
10. Sigurdsson, H.B.; Maguire, M.C.; Balascio, P.; Silbernagel, K.G. Effects of kinesiophobia and pain on performance and willingness to perform jumping tests in Achilles tendinopathy: A cross-sectional study. *Phys. Ther. Sport* **2021**, *50*, 139–144. [CrossRef] [PubMed]
11. Novacheck, T.F. The biomechanics of running. *Gait Posture* **1998**, *7*, 77–95. [CrossRef]
12. Morin, J.B.; Samozino, P.; Millet, G.Y. Changes in running kinematics, kinetics, and spring-mass behavior over a 24-h run. *Med. Sci. Sports Exerc.* **2011**, *43*, 829–836. [CrossRef]
13. Chollet, M.; Michelet, S.; Horvais, N.; Pavailler, S.; Giandolini, M. Individual physiological responses to changes in shoe bending stiffness: A cluster analysis study on 96 runners. *Eur. J. Appl. Physiol.* **2023**, *123*, 169–177. [CrossRef]
14. McMahon, T.A. Spring-like properties of muscles and reflexes in running. In *Multiple Muscle Systems: Biomechanics and Movement Organization*; Springer: New York, NY, USA, 1990; pp. 578–590. [CrossRef]

15. Lin, S.P.; Sung, W.H.; Kuo, F.C.; Kuo, T.B.; Chen, J.J. Impact of center-of-mass acceleration on the performance of ultramaratho runners. *J. Hum. Kinet.* **2014**, *44*, 41. [CrossRef]
16. Lee, C.R.; Farley, C.T. Determinants of the center of mass trajectory in human walking and running. *J. Exp. Biol.* **1998**, *201* 2935–2944. [CrossRef] [PubMed]
17. Möhler, F.; Stetter, B.; Müller, H.; Stein, T. Stride-to-stride variability of the center of mass in male trained runners after ar exhaustive run: A three-dimensional movement variability analysis with a subject-specific anthropometric model. *Front. Sport. Act. Living* **2021**, *3*, 665500. [CrossRef]
18. Jaén-Carrillo, D.; Roche-Seruendo, L.E.; Felton, L.; Cartón-Llorente, A.; García-Pinillos, F. Stiffness in running: A narrativ integrative review. *Strength. Cond. J.* **2021**, *43*, 104–115. [CrossRef]
19. Latash, M.L.; Zatsiorsky, V.M. Joint stiffness: Myth or reality? Hum. *Mov. Sci.* **1993**, *12*, 653–692. [CrossRef]
20. McMahon, T.A.; Cheng, G.C. The mechanics of running: How does stiffness couple with speed? *J. Biomech.* **1990**, *23*, 65–78 [CrossRef]
21. Hennig, E.M.; Lafortune, M.A. Relationships between ground reaction force and tibial bone acceleration parameters. *J. Appl Biomech.* **1991**, *7*, 303–309. [CrossRef]
22. Bishop, M.; Fiolkowski, P.; Conrad, B.; Brunt, D.; Horodyski, M. Athletic footwear, leg stiffness, and running kinematics. *J. Ath Train.* **2006**, *41*, 387.
23. Moore, I.S.; Ashford, K.J.; Cross, C.; Hope, J.; Jones, H.S.; McCarthy-Ryan, M. Humans optimize ground contact time and leg stiffness to minimize the metabolic cost of running. *Front. Sports Act. Living* **2019**, *1*, 53. [CrossRef]
24. McMahon, T.A.; Valiant, G.; Frederick, E.C. Groucho running. *J. Appl. Physiol.* **1987**, *62*, 2326–2337. [CrossRef]
25. Van Hooren, B.; Jukic, I.; Cox, M.; Frenken, K.G.; Bautista, I.; Moore, I.S. The relationship between running biomechanics and running economy: A systematic review and meta-analysis of observational studies. *Sports Med.* **2024**, *54*, 1269–1316. [CrossRef
26. Skime, A.; Boone, T. Cardiovascular responses during Groucho running. *J. Exerc. Physiol. Online* **2011**, *14*, 88–92.
27. Lussiana, T.; Gindre, C.; Hébert-Losier, K.; Sagawa, Y.; Gimenez, P.; Mourot, L. Similar running economy with different running patterns along the aerial-terrestrial continuum. *Int. J. Sports Physiol. Perform.* **2017**, *12*, 481–489. [CrossRef] [PubMed]
28. Walker, C.; Blair, R. An experimental review of the McMahon/Cheng model of running. *Sports Eng.* **2001**, *4*, 113–121. [CrossRef
29. Bertram, J.E.; D'antonio, P.; Pardo, J.; Lee, D.V. Pace length effects in human walking: "Groucho" gaits revisited. *J. Mot. Behav* **2002**, *34*, 309–318. [CrossRef]
30. Kung, S.M.; Fink, P.W.; Legg, S.J.; Ali, A.; Shultz, S.P. What factors determine the preferred gait transition speed in humans? A review of the triggering mechanisms. *Hum. Mov. Sci.* **2018**, *57*, 1–12. [CrossRef] [PubMed]
31. Bailey, J.P.; Silvernail, J.F.; Dufek, J.S.; Navalta, J.; Mercer, J.A. Effects of treadmill running velocity on lower extremity coordination variability in healthy runners. *Hum. Mov. Sci.* **2018**, *61*, 144–150. [CrossRef]
32. Monteiro, A.S.; Galano, J.P.; Cardoso, F.; Buzzachera, C.F.; Vilas-Boas, J.P.; Fernandes, R.J. Kinematical and physiological responses of overground running gait pattern at different intensities. *Sensors* **2024**, *24*, 7526. [CrossRef]
33. Patoz, A.; Lussiana, T.; Breine, B.; Gindre, C.; Malatesta, D.; Hébert-Losier, K. Examination of running pattern consistency across speeds. *Sports Biomech.* **2025**, *24*, 200–214. [CrossRef]
34. Farley, C.T.; Gonzalez, O. Leg stiffness and stride frequency in human running. *J. Biomech.* **1996**, *29*, 181–186. [CrossRef]
35. Cavagna, G.A. Force platforms as ergometers. *J. Appl. Physiol.* **1975**, *39*, 174–179. [CrossRef] [PubMed]
36. Koo, T.K.; Li, M.Y. A guideline of selecting and reporting intraclass correlation coefficients for reliability research. *J. Chiropr. Med* **2016**, *15*, 155–163. [CrossRef]
37. Cohen, J. *Statistical Power Analysis for the Behavioral Sciences*; Routledge: Oxfordshire, UK, 2013. [CrossRef]
38. Gindre, C.; Lussiana, T.; Hebert-Losier, K.; Mourot, L. Aerial and terrestrial patterns: A novel approach to analyzing human running. *Int. J. Sports Med.* **2016**, *37*, 25–26. [CrossRef]
39. Smith, C.P.; Fullerton, E.; Walton, L.; Funnell, E.; Pantazis, D.; Lugo, H. The validity and reliability of wearable devices for the measurement of vertical oscillation for running. *PLoS ONE* **2022**, *17*, e0277810. [CrossRef]
40. Gouttebarge, V.; Wolfard, R.; Griek, N.; de Ruiter, C.J.; Boschman, J.S.; van Dieën, J.H. Reproducibility and validity of the myotest for measuring step frequency and ground contact time in recreational runners. *J. Hum. Kinet.* **2015**, *45*, 19–26. [CrossRef] [PubMed]
41. Webber, E.; Leduc, C.; Emmonds, S.; Eglon, M.; Hanley, B.; Iqbal, Z.; Sheoran, S.; Chaisson, C.; Weaving, D. From lab to field: Validity and reliability of inertial measurement unit-derived gait parameters during a standardised run. *J. Sports Sci.* **2024**, *42* 1706–1715. [CrossRef] [PubMed]

biomechanics

MDPI

rticle

Fatigue-Related Biomechanical Changes During a Half-Marathon Under Field Conditions Assessed Using Inertial Measurement Units

Christian Mitschke [1,*], Tobias Heß [1,2], Thomas L. Milani [2] and Pierre Kiesewetter [2]

1 Department of Movement Science for Prevention and Rehabilitation, Institute of Human Movement Science and Health, Chemnitz University of Technology, Thüringer Weg 11, 09126 Chemnitz, Germany; tobias.hess@hsw.tu-chemnitz.de

2 Research Group for Biomechanics and Sensory Function, Institute of Human Movement Science and Health, Chemnitz University of Technology, Thüringer Weg 11, 09126 Chemnitz, Germany; thomas.milani@hsw.tu-chemnitz.de (T.L.M.); pierre.kiesewetter@hsw.tu-chemnitz.de (P.K.)

* Correspondence: christian.mitschke@hsw.tu-chemnitz.de; Tel.: +49-371-531-32196

Abstract: Background/Objectives: Running is one of the most popular physical activities worldwide and have been widely studied in relation to performance and injury prevention. In addition to measurements conducted under standardized laboratory conditions, inertial measurement units (IMUs) allow for the assessment of biomechanical parameters in real-world settings—particularly during endurance runs. The aim of this study was to investigate how running a half-marathon under field conditions affects exertion and various biomechanical parameters, as measured using IMUs. **Methods**: Twenty runners completed a half-marathon on a flat, even-surfaced walkway at a self-selected, constant pace corresponding to a brisk training run. In addition to lower limb biomechanics, heart rate (HR) and ratings of perceived exertion (REP) were also recorded. **Results**: A significant increase in both HR and RPE was observed toward the end of the half-marathon, indicating the presence of fatigue during the later stages of the run. The biomechanical results further demonstrate that this fatigue was associated with increased peak tibial acceleration, peak angular velocity in the sagittal plane of the foot, and peak rearfoot eversion velocity, while foot strike angle, stride frequency, and stride length remained unchanged. Furthermore, a progressive increase in ground contact time and a decrease in flight time were observed over the course of the run, resulting in an increased duty factor. **Conclusions**: These findings highlight the value of IMU-based assessments for detecting fatigue-related biomechanical changes during prolonged runs in real-world conditions, which may contribute to early identification of overload and inform injury prevention strategies.

Keywords: running; half-marathon; biomechanics; inertial measurement unit; peak tibial acceleration; peak rearfoot eversion velocity

1. Introduction

Running is one of the most widely practiced forms of physical activity and is associated with substantial health benefits. Beyond its health-promoting effects, the appeal of competition and performance comparison has become increasingly prominent, as evidenced by the growing number of marathon or half-marathon participants worldwide [1–3]. However, running also carries a notably high incidence of overuse injuries, with prevalence ranging from 19.4% to 92.4% among distance runners [4]. This is likely attributable in part to increased training volumes and intensities during competition preparation phases, although the relationship between training load and injury risk remains a subject of ongoing debate [5,6].

Given the multifactorial etiology of overuse injuries, attention has increasingly turned to biomechanical parameters that may act as modifiable risk factors.

In this context, several studies have focused on biomechanical parameters in distance running particularly in relation to muscular fatigue, which naturally accumulates during prolonged endurance runs or at higher intensities, in healthy, recovered, and acutely symptomatic runners [7–23].

One of these biomechanical parameters is peak tibial acceleration (PTA), which provides information on how the shock wave resulting from the forces at initial ground contact (IC) propagates through the lower extremities and allows inferences about the internal forces acting on the musculoskeletal system [24,25]. Thereby, excessive values in PTA have been associated with increased mechanical strain on the tibia and surrounding structures, potentially predisposing athletes to bone stress injuries [11,26–30]. The influence of muscular fatigue on PTA has been examined primarily during running on treadmills or indoor running tracks, with inconsistent results. While some investigations suggest that muscular fatigue induces adaptations in running mechanics—reflected in increased PTA values [9,11]—other studies have not reported consistent evidence of such changes [7,12]. To date, apart from indoor running studies, only a few investigations have addressed PTA during endurance running in real-world settings, likewise with inconsistent results. For example, Morio et al. [18] reported no change in PTA during field running, whereas Ruder et al. [22] observed a decrease at the end of an endurance run.

In addition to the impact loading parameter PTA, it is also relevant to consider the influence of muscular fatigue on kinematic measures such as peak eversion velocity (evVel), peak angular velocity in the sagittal plane (PAV), and foot strike angle (FSA). In this context, several studies have shown that with increasing muscular fatigue, evVel increases significantly [10,11,31], whereas FSA may decrease significantly [12,20,32]. No recent study could be identified that examined the influence of muscular fatigue on PAV. However, this parameter is highly relevant as it reflects the foot rollover in the sagittal plane [33]. It may provide insights into the ability to actively decelerate plantarflexion after IC, a function that relies on the dorsiflexors and may be impaired by their fatigue, which in turn can influence shock absorption [34].

Furthermore, fatigue-related changes in spatiotemporal parameters—including step frequency (SF), ground contact time (t_C), flight time (t_F), and duty factor (DF)—are of particular interest, as they represent fundamental descriptors of running mechanics. They are generally stable at constant speeds but may show subtle changes under muscular fatigue, which can reflect compensatory strategies and influence running efficiency or injury risk [7–9,11,12,18]. However, the existing literature reports inconsistent findings regarding these adaptations, which may be partly explained by differences in methodological approaches such as running speed, exercise intensity, or running surface.

As shown, most studies investigating the effects of muscular fatigue on the abovementioned biomechanical parameters have been conducted on treadmills or indoor running tracks, often with inconsistent results [7–11,15,17]. Further investigations have been carried out under field conditions, likewise reporting inconsistent findings or not including relevant parameters of impact loading (e.g., PTA) and foot rollover kinematics (e.g., evVel) in conjunction with spatiotemporal parameters [13,14,19–21,23].

These inconsistencies point to a broader issue: muscular fatigue appears to elicit different biomechanical responses depending on the running context. Specifically, the existing literature indicates that biomechanical adaptations in response to muscular fatigue may differ between treadmill and field running. This suggests that changes in parameters such as PTA are strongly influenced by the study design—particularly the running surface, speed, and intensity. Supporting this, higher PTA values have been reported during outdoor running compared to treadmill conditions [35–37]. Consequently, biomechanical variables associated with overuse injuries—such as medial tibial stress syndrome (MTSS) or tibial stress fractures (TSF)—identified through laboratory-based gait analysis may not accurately reflect their magnitudes during field running. To address this limitation, studies conducted in natural environments (e.g., on streets or sidewalks) are necessary to investigate how muscular fatigue influences biomechanical parameters under ecologically valid conditions [38]. Furthermore, a review by Xiang et al. [39] highlighted that, beyond impact parameters such as PTA or ground reaction forces, a holistic approach should be adopted by integrating additional biomechanical aspects, with inertial measurement units (IMUs) providing a promising tool for this purpose.

In this context, IMUs, which combine accelerometers and gyroscopes, offer a particularly suitable solution for field-based research due to their compact size and lightweight design, as has been demonstrated in several field-based studies [13,18,33,36,37].

As demonstrated above, previous findings remain inconsistent, underscoring a crucial gap in knowledge regarding how fatigue affects running biomechanics under field conditions. This highlights the need for approaches that enable the comprehensive assessment of multiple biomechanical parameters outside the laboratory. Therefore, the aims of this study were twofold: (1) to investigate whether IMUs are suitable for capturing a broad range of biomechanical parameters during a half-marathon run under field conditions, extending beyond the more limited approaches used in previous research, and (2) to examine how such a run affects exertion and the above mentioned biomechanical parameters. It was hypothesized (I) that both heart rate and perceived exertion would significantly increase with the duration of the run. With respect to biomechanics, it was hypothesized (II) that PTA would significantly increase from the beginning to the end of the run as a result of acute muscular fatigue and a reduced capacity to absorb impact forces. Furthermore, it was hypothesized (III) that foot kinematics—specifically, evVel and PAV—would increase over the course of the run due to progressive muscular fatigue in the lower extremities, whereas the FSA would decrease. Regarding spatiotemporal parameters, it was hypothesized (IV) that, despite a constant running speed, increasing fatigue during a half marathon would lead to a decrease in strLen and t_F, while SF and t_C would increase. We also hypothesized (V) that exertion parameters are correlated with biomechanical parameters.

This study aims to demonstrate the potential of IMU-based systems for continuous, field-based monitoring of running biomechanics under fatigue conditions during prolonged endurance runs. Understanding how biomechanical parameters change under real-world conditions provides valuable insights for injury prevention in endurance runners. In addition, the findings may inform training strategies and fatigue monitoring through real-time feedback to promote safer and more efficient running over longer distances.

2. Materials and Methods

2.1. Participants

In summary, 20 runners without any injuries in the last six months were recruited for this study. Demographic and running-related characteristics of the runners are presented in Table 1. Participants ran an average of 36.8 ± 21.8 km per week. During the half-marathon run, the average running speed was 11.4 ± 0.9 km/h.

Table 1. Demographic and running-related characteristics of the runners (mean ± SD).

Age [Years]	Height [cm]	Weight [kg]	Gender [Male/Female]	Weekly Running Distance [km]	Running Speed [km/h]
36.9 ± 12.4	177.2 ± 6.4	71.5 ± 10.8	16/4	36.8 ± 21.8	11.4 ± 0.9

The runners were informed about the purpose and design of this study, signed an informed consent document, and completed a form with their personalized data. All procedures were performed in accordance with the recommendations of the Declaration of Helsinki. This study was approved by the Ethics Committee of the Chemnitz University of Technology (#101627696).

2.2. Experimental Setup and Procedures

2.2.1. Running

Data acquisition took place on a straight, flat concrete sidewalk approximately 1.4 km in length. Following an individual warm-up period of 5 to 10 min, each participant completed 15 laps on the sidewalk, totaling the official half-marathon distance of approximately 21.1 km. Running speed was self-selected and corresponded to each runner's typical training pace, as suggested in previous studies [10,18]. Participants were instructed to choose a running pace they could maintain continuously over the entire half-marathon distance without the need to stop; however, a minimum speed of 10 km/h was required. The running speed was constantly checked by the experimenter on a bicycle with a speedometer [33,36]. During the half-marathon run, participants used their own running shoes. Runners rated their perceived exertion (RPE) on a 15-point Borg scale (from 6 to 20) at the beginning of the run and after each lap [40]. Furthermore, participants wore a Garmin Forerunner® 735XT (Garmin, Olathe, KS, USA) equipped with an external heart rate sensor (HRM-Pro™ Plus, Garmin, Olathe, KS, USA) to monitor heart rate (HR).

2.2.2. Sensor Setup

Based on Hill et al. [36], four small, lightweight inertial measurement units (IMU; ICM-20601, InvenSense, San Jose, CA, USA; weight: 4 g; sampling rate: 2000 Hz) combining a tri-axial accelerometer (measurement range: ± 353 m/s^2) and a tri-axial gyroscope (measurement range: $\pm 4000°$/s) were used to measure biomechanical data (Figure 1A). All IMUs were connected via cables to a data logger, which was secured on a belt around the participant's waist. According to Kiesewetter et al. [41], one IMU was attached to the shaved medial aspect of each tibia, midway between the malleolus and tibial plateau, using double-sided adhesive tape, an elastic strap, and a compression sleeve to minimize sensor movement. The sensor axes were aligned with the longitudinal axes of the tibia. Additionally, an IMU was mounted on the heel cap of each running shoe using double-sided adhesive tape and secured with additional elastic tape.

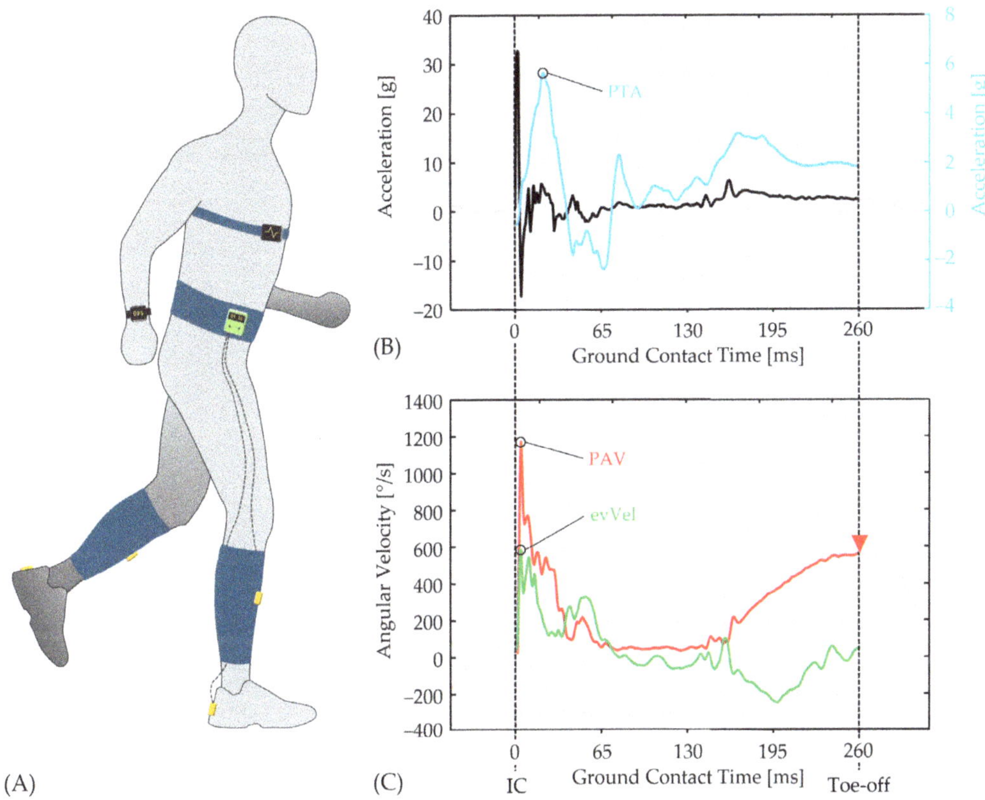

Figure 1. (**A**) Schematic illustration of the setup with inertial measurement units (IMUs), data logger, Garmin Forerunner®, and heart rate sensor; (**B**) Vertical acceleration data from the heel-mounted IMU (black line) as well as from the tibia-mounted IMU (blue line), showing peak tibial acceleration (PTA); (**C**) Angular velocities from the heel-mounted IMU in the frontal plane (green line) and sagittal plane (red line), used to determine peak rearfoot eversion velocity (evVel) and peak angular velocity for foot rollover sagittal (PAV). Toe-off was defined as maximum angular velocity in the sagittal plane, >100 ms post IC (red triangle). A sample dataset was used for illustration.

2.3. Data Analysis

Raw data from the IMUs were analyzed post-processing using MATLAB 2024a (MathWorks™, Natick, MA, USA). Prior to analysis, signals were filtered to reduce noise using a zero-lag, 4th-order Butterworth low-pass filter—applied at 80 Hz for accelerometer data and 50 Hz for gyroscope data [41].

Biomechanical Parameters

For accurate detection of IC and stride segmentation, the vertical acceleration signal from the heel-mounted IMU was further processed using a zero-lag Butterworth high-pass filter at 80 Hz. The first prominent peak in the filtered signal was identified as IC, ensuring a valid detection of foot strike events for subsequent gait analysis [42]. The stride duration (strD) was then calculated as the time between two consecutive ICs. In addition, the SF was calculated as the inverse of the time between two consecutive ICs, multiplied by two (assuming that the right and left legs have the same step duration). According to Sabatini et al. [43], toe-off was identified as the point at which the angular velocity in the sagittal plane of the foot-mounted sensor reaches its maximum value, occurring at least 100 ms after IC (Figure 1C). Furthermore, t_C was defined as the interval between IC and the corresponding toe-off event, while t_F was computed as follows (1):

$$t_F = strD - t_C \quad (1)$$

The duty factor (DF), representing the proportion of time the foot is in contact with the ground during a stride, was used to assess mechanical work [20,44]. A higher DF indicates a longer t_C, whereas a lower DF reflects longer t_F and reduced t_C. The following Equation (2) was used to calculate the relative DF:

$$DF = t_C\ (t_F + t_C)^{-1} * 100\% \quad (2)$$

The highest peak in the vertical acceleration signal of the tibia-mounted sensor was defined as PTA (Figure 1B). To assess evVel, the maximum angular velocity in the frontal plane of the shoe-mounted sensor was analyzed [45] (Figure 1C). For foot rollover, PAV was determined within 100 ms after IC, following the approach described by Bräuer et al. [33] (Figure 1C). To determine the FSA at IC, the orientation of the shoe in the sagittal plane was calculated using angular velocity data from the heel-mounted IMU [46]. This signal was integrated to obtain the foot angle (θ). To correct for integration drift and offset, a nulling algorithm was applied using two consecutive stance phases where the foot was flat on the ground, identified by minimal changes in θ. A linear offset correction between these points was subtracted from θ. FSA was then defined as the corrected orientation angle at IC. A detailed description of this procedure can be found in Mitschke et al. [46]. The drift- and offset-corrected θ signal, along with the corresponding vertical and horizontal acceleration data, was used to calculate strLen between two consecutive ICs [46]. Due to technical issues (such as cable breakage or sensor failures) occurring during testing, all biomechanical parameters were calculated based on data from only one leg per runner.

2.4. Statistics

Each runner completed a total of 15 laps. For each lap, means were calculated for all biomechanical parameters. Subsequently, for lap 1 as well as laps 8 and 15, group means and standard deviations (mean $\pm$ SD) were calculated across all runners. The three laps represent different race segments: the start (L1), running in a "non-fatigued state", running in the middle of the race (L8), and running in a "fatigued state" (L15). This segmentation into three phases was based on the approach of Prigent et al. [20].

Statistical analyses were performed using IBM SPSS Statistics (IBM Corp., Armonk, NY, USA, Version 30.0). All data were visually inspected and tested for normal distribution using the Shapiro–Wilk test. In cases where normality was confirmed, a one-way repeated-measures ANOVA was conducted to compare the three laps (L1, L8, and L15), followed by a Bonferroni-corrected post hoc test. If the assumption of normality was violated, the Friedman test was applied for the same laps, with subsequent Dunn-Bonferroni tests used for post hoc analysis. Statistical significance was set at $\alpha = 0.05$ for all analyses. In addition, effect size (Cohen's d) was calculated to quantify the magnitude of differences when statistical significance was found. The coefficients were interpreted as trivial ($d < 0.2$), small ($d < 0.5$), medium ($d < 0.8$), or large effects ($d \geq 0.8$) [47]. Pearson's correlation coefficient was used to assess linear relationships between exertion parameters (RPE and HR) and biomechanical parameters.

3. Results

3.1. Exertion Parameters

Table 2 presents the results for the exertion parameters. For both parameters, the lowest values were observed in L1 and the highest in L15. The ANOVA revealed significant differences across laps. Subsequent post hoc tests showed significant pairwise differences between all laps, with large effect sizes (d > 0.98).

Table 2. Effects of running duration and muscular fatigue on heart rate (HR) and perceived exertion (RPE) (mean ± SD). Statistically significant group differences are indicated with *, while significant differences between laps 1 (L1), 8 (L8), and 15 (L15) are marked with a, b, and c, respectively; effect sizes (Cohen's d) are reported for the respective pairwise comparisons.

	L1	L8	L15	*p*-Value	d
HR [1/min]	133.8 ± 13.8 a; b	148.4 ± 15.8 a; c	156.6 ± 19.2 b; c	* < 0.001	
				a < 0.001	a = 0.98
				b < 0.001	b = 1.36
				c < 0.001	c = 0.47
RPE (6–20)	9.0 ± 1.6 a; b	12.8 ± 2.3 a; c	15.9 ± 2.2 b; c	* < 0.001	
				a < 0.001	a = 1.90
				b < 0.001	b = 3.53
				c < 0.001	c = 1.36

3.2. Biomechanical Parameters

The biomechanical parameters for the three examined segments of the half-marathon are presented in Table 3 and Figure 2. An increase in PTA over the course of the run was measured. ANOVA revealed significant differences between L1 and L15 ($p < 0.001$) with a medium effect size (d = 0.58). For both evVel and PAV, increases in angular velocities were observed from L1 to L15. While evVel showed only a trend toward higher velocities, PAV exhibited significant differences between all three laps (L1 vs. L8: $p = 0.046$; L1 vs. L15: $p = 0.011$; L8 vs. L15: $p = 0.024$), with trivial (L8 vs. L15: d = 0.012) to small effect sizes (L1 vs. L8: d = 0.31; L1 vs. L15: d = 0.42). No significant difference in FSA was found between the laps ($p = 0.210$).

Table 3. Effects of running duration and muscular fatigue on peak tibial acceleration (PTA), peak eversion velocity (evVel), peak angular velocity in the sagittal plane (PAV), and foot strike angle (FSA) (mean ± SD). Statistically significant group differences are indicated with *, while significant differences between laps 1 (L1), 8 (L8), and 15 (L15) are marked with a, b, and c, respectively; effect sizes (Cohen's d) are reported for the respective pairwise comparisons.

	L1	L8	L15	*p*-Value	d
PTA [g]	8.7 ± 3.5 a	9.8 ± 4.2	10.1 ± 4.2 a	* < 0.001	
				a < 0.001	a = 0.58
evVel [°/s]	425.1 ± 183.0	440.4 ± 160.0	452.1 ± 154.8	0.169	-
PAV [°/s]	854.2 ± 135.3 a; b	900.6 ± 160.4 a; c	919.9 ± 172.0 b; c	* 0.004	
				a = 0.046	a = 0.31
				b = 0.011	b = 0.42
				c = 0.024	c = 0.12
FSA [°]	24.9 ± 3.7	25.4 ± 3.7	24.7 ± 4.1	0.210	-

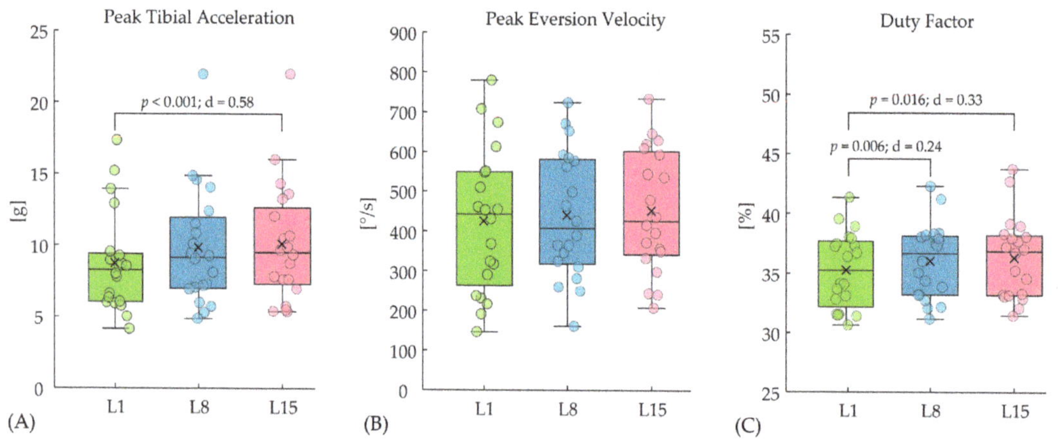

Figure 2. Comparison of the three examined laps of the half-marathon (L1, L8, and L15) for the biomechanical parameters (**A**) peak tibial acceleration, (**B**) peak eversion velocity, and (**C**) duty factor.

Table 4 presents the results for spatiotemporal parameters. No significant differences were found for strLen ($p = 0.484$) and SF ($p = 0.210$). Moreover, neither parameter showed an increasing or decreasing trend over the three laps. For t_C, a significant increase with a small effect size was observed between laps L1 and L8 ($p = 0.022$; d = 0.29) as well as between L1 and L15 ($p = 0.044$; d = 0.34). In contrast, t_F significantly decreased over the course of the run from lap L1 to L8 ($p = 0.036$; d = 0.15) and from L1 to L15 ($p = 0.020$; d = 0.23). For DF, ANOVA revealed significant differences between L1 and L8 ($p = 0.006$) as well as between L1 and L15 ($p = 0.016$), both indicating small effect sizes (L1 vs. L8: d = 0.24; L1 vs. L15: d = 0.33) (Figure 2C).

Table 4. Effects of running duration and muscular fatigue on spatiotemporal parameters stride length (strLen), step frequency (SF), ground contact time (t_C), flight time (t_F), and duty factor (DF) (mean ± SD). Statistically significant group differences are indicated with *, while significant differences between laps 1 (L1), 8 (L8), and 15 (L15) are marked with a and b, respectively; effect sizes (Cohen's d) are reported for the respective pairwise comparisons.

	L1	**L8**	**L15**	***p*-Value**	**d**
strLen [cm]	206.9 ± 22.6	207.0 ± 22.0	206.2 ± 22.0	0.484	-
SF [1/min]	166.2 ± 9.6	166.6 ± 9.3	167.1 ± 9.1	0.210	-
t_C [ms]	255.2 ± 17.8 a; b	260.4 ± 17.6 a	261.4 ± 19.3 b	* 0.010	
				a = 0.022	a = 0.29
				b = 0.044	b = 0.34
t_F [ms]	469.4 ± 45.0 a; b	462.8 ± 43.2 a	459.0 ± 44.1 b	* 0.006	
				a = 0.036	a = 0.15
				b = 0.020	b = 0.23
DF [%]	35.3 ± 3.2 a; b	36.1 ± 3.1 a	36.4 ± 3.4 b	* 0.004	
				a = 0.006	a = 0.24
				b = 0.016	b = 0.33

Pearson correlation coefficients were calculated to investigate the relationships between exertion parameters and biomechanics. Moderate positive correlations were found between evVel and both RPE (ΔL15–L1: r = 0.49, $p = 0.030$) and HR (ΔL15–L1: r = 0.45,

$p = 0.045$). Other biomechanical parameters showed very low correlation coefficients with RPE and HR, none of which reached statistical significance.

4. Discussion

Previous studies have shown that muscular fatigue can substantially influence running biomechanics, which has been associated with performance reduction and a potentially elevated risk of injury [48]. In the present study, muscular fatigue was induced by having participants run a half-marathon. By using of IMUs, biomechanical data could be collected under real-world conditions, which resulted in findings with higher ecological validity. For the half-marathon run, the running speed was set individually for each participant to correspond to a brisk training run but was limited to a speed that allowed the run to be completed without interruption. Maintaining a constant running speed throughout the entire run was essential to avoid speed-related confounding effects on parameters such as PTA, strLen, SF, and others. It was hypothesized that (I) exertion parameters (II) impact loading, (III) kinematics, and (IV) spatiotemporal parameters change as fatigue progressed. Furthermore, we hypothesized that there is a correlation between exertion parameters and biomechanical parameters (V).

4.1. Exertion Parameters

During the half-marathon run, a significant increase in fatigue was observed from L1 to L8 and L15, as indicated by raised exertion parameters HR and RPE, both demonstrating large effect sizes (Table 2). Therefore, it can be assumed that the runners experienced a high level of physiological fatigue, particularly towards the end of the run. This assumption is supported by significant fatigue-related changes in several biomechanical parameters. Based on the significant increase and the large effect in exertion parameters, Hypothesis (I) can be confirmed.

4.2. Impact Loading

At each foot strike, especially at IC, the lower extremities experience impact forces several times body weight within a short time [49,50]. During prolonged runs, this mechanical loading contributes to muscular fatigue, which reduces the ability of the muscles to effectively attenuate impact forces [8]. In this context, Mizrahi et al. [34] demonstrated that muscle fatigue of the shank, including the tibialis anterior, was associated with a reduction in the mean power frequency of electromyographic activity—an established indicator of muscular fatigue. At the same time, their study reported a significant increase in PTA. It can be concluded that such muscular fatigue results in a diminished or delayed shock-attenuation response. Our study revealed a significant increase in PTA throughout the run, corresponding to a medium effect size (Table 3). This indicates that increased muscular fatigue in the lower extremities was pronounced in our study, making it more difficult for the runners to effectively absorb and manage shock waves generated during each foot strike. Therefore, our findings provide support for confirming Hypothesis (II).

Our findings are consistent with previous studies conducted on treadmills [9,11]. For example, Derrick et al. [11] investigated PTA during an exhaustive treadmill run. Muscular fatigue resulted in a significant 20.7% increase in PTA at the end of the run. While PTA values may be lower due to the treadmill setting, where reduced maxima are expected compared to sidewalk running [35–37], the fatigue-induced increase remains comparable to that observed in our study. Furthermore, Mizrahi et al. [9] demonstrated that fatigue during a 30 min treadmill run led to an approximately 50% increase in PTA, suggesting that fatigue significantly amplifies mechanical loading on the shank and may elevate the risk of overload injuries. The higher PTA increase reported in the referenced study was probably due to the intensity being above the anaerobic threshold, in contrast to our study, where participants should run at lower intensity (below anaerobic threshold) for a longer duration.

Similarly to our study, Morio et al. [18] investigated PTA during a continuous run conducted on a road. However, in contrast to our findings, the authors did not observe any significant changes in PTA by the end of the running session. In their study, participants ran at their self-selected constant speed, similar to a typical 1 h training run. Apparently, the duration of the run (1 h) or the running speed was too low and thus the level of fatigue was insufficient to elicit a significant increase in PTA. The authors explained the low RPE values as an indication that the exercise intensity was not high enough to induce substantial fatigue [18]. In contrast, Ruder et al. [22] reported a pronounced level of fatigue in their participants, who completed a full marathon. Toward the end of the run, a significantly reduced PTA was observed. The authors attributed the approximately 15% decrease in PTA to a corresponding 15% reduction in running speed at the end of the race, likely resulting from fatigue-induced performance decline. Subsequently, the authors normalized PTA to running speed and showed that, after adjusting for speed, PTA remained unchanged toward the end of the run. This observation further highlights the contrast to the results of the present study, where PTA increased despite a constant running speed.

4.3. Kinematics

Regarding kinematics, only PAV increased significantly with running duration (Table 3). This change likely reflects fatigue-induced alterations in neuromuscular control, particularly involving the tibialis anterior. The ability to actively decelerate the plantarflexion after IC appeared to diminish, likely due to compromised eccentric control by the dorsiflexors, leading to progressively higher angular velocities across the analyzed segments. The PAV values observed in this study are in line with previously reported findings, such as those examining the influence of running shoes on biomechanics (757.8–814.1°/s) [41] or assessing foot rollover characteristics (845.6–919.5°/s) [33], when measured using IMUs. However, no comparable values for PAV during prolonged runs could be found in the existing literature.

Notably, despite the significant increase in PAV, the FSA remained unchanged throughout the run, suggesting that runners maintained a consistent foot strike pattern despite muscular fatigue (Table 3). FSA values are comparable to those reported in the literature [41,51,52].

To date, only a limited number of studies have examined how muscular fatigue affects FSA. For example, Prigent et al. [20] reported a significant reduction in FSA during a road half-marathon run. This reduction suggests a flatter foot placement at IC. During treadmill running, Hazzaa and Mattes [32] observed a significant reduction in FSA following muscular fatigue, particularly when the plantarflexors were fatigued. Furthermore, Giandolini et al. [12] reported that a 110 km ultramarathon induced significant FSA adaptations depending on foot strike pattern: non-rearfoot strikers showed an increase, while rearfoot strikers showed a decrease, resulting in convergence between groups. In contrast, other studies have reported no change in FSA with increasing fatigue. For example, Camelio et al. [7] observed no change in FSA before and after a 30 min submaximal treadmill run. The authors attributed this to the low intensity (approx. 12/20 on RPE scale), which may not have been sufficient to induce muscular fatigue affecting FSA.

In summary, FSA can be influenced by muscular fatigue when the level of exertion is sufficiently high, and this effect appears to depend on the foot strike pattern. In our study, all runners exhibited FSA > 15° at baseline, indicating a consistent rearfoot striking; thus, further subdivision was unnecessary. Although exertion increased substantially (RPE: 15.9 ± 2.2), no significant FSA changes were observed. Even when participants were subdivided into higher- and lower-fatigue groups (RPE > 16 or HR > 165 bpm), results for FSA remained inconsistent.

Overall, the findings do not provide consistent evidence that localized muscle fatigue induces measurable changes in FSA. Comparable results were found for evVel. Although a slight increasing trend was observed across all runners on average, the change was not statistically significant (Table 3). Generally, similar values for evVel have been reported in other studies [31,41,45]. Only few studies have investigated the effects of muscular fatigue on evVel. For example, Shih et al. [31] reported a significant increase during an intense 30 min treadmill run, with strong correlations to RPE and HR (r = 0.911 and r = 0.960). Our finding showed similar associations between evVel and exertion, though with moderate correlation strength. These results suggest that increasing exertion is linked to higher evVel, potentially reflecting fatigue-related adaptations in lower limb control, even though evVel did not significantly change across laps.

Derrick et al. [11] also observed a significant increase in evVel during an exhaustive treadmill run. The relative increase (12.6%) was larger than in our study (6.4%), which may be explained by their higher running speeds, as evVel is known to rise with speed (12.2 km/h vs. 11.4 km/h) [45]. In contrast, Dierks et al. [10] reported markedly lower evVel in runners with and without patellofemoral pain (PFP), likely due to slower running speeds (~9 km/h). In both groups, evVel increased significantly during the run, with similar relative changes, suggesting no specific effect of PFP on evVel. Generally, the current evidence regarding biomechanical risk factors for running-related injuries remains conflicting [5]. In particular, there is limited evidence regarding the relationship between muscular fatigue and running-related injuries.

As our results indicate, a significant effect of muscular fatigue was observed only for PAV. For evVel, a trend toward higher values at the end of the run was found, as also reflected in the correlations between evVel and HR as well as rating of RPE. In contrast, no fatigue-related effect was evident for the FSA. Therefore, Hypothesis (III) can only be partially accepted.

4.4. Spatiotemporal Parameters

Regarding spatiotemporal parameters, the results of this study indicate that both SF and strLen remained constant throughout the half-marathon, despite progressive muscular fatigue (Table 3). This suggests that running speed was maintained consistent across the entire race.

The influence of fatigue on spatiotemporal parameters has been reported controversially in the literature. Our results on SF and strLen align with previous studies showing stable spatiotemporal parameters at consistent running speeds [7,11]. In contrast, other studies have reported that muscular fatigue leads to an increase in SF, which is often accompanied by a reduction in strLen [9,12,18]. SF has been proposed to increase as a compensatory strategy to reduce fatigue-related impact forces [18], serving as a protective mechanism against musculoskeletal loading. In this context, Verbitsky et al. [8] reported a decrease in SF in fatigued runners after a 30 min treadmill run near the anaerobic threshold, while it remained stable in a non-fatigue group. These findings, together with the absence of significant SF or strLen changes in our study and others (e.g., [7]), suggest that participants may not have reached sufficient fatigue to alter these parameters. However, Derrick et al. [11] reported very similar findings to our study during an exhaustive run.

The authors observed an increase in PTA towards the end of the run, while both strLen and SF remained unchanged. Furthermore, the runners in their study exhibited a lower knee angle (higher flexion), a higher evVel as well as an increased rearfoot eversion angle at IC—findings that are consistent with the results of evVel observed in our study. According to the authors, changes in knee and subtalar joint angles at IC likely reduced the effective mass of the lower limb during impact, which in turn may have contributed to the observed increases in leg impact accelerations and shock attenuation. The authors emphasize that an increase in measured PTA does not necessarily indicate a higher injury risk if it results from a reduction in effective mass [11]. Since it is primarily the impact forces—rather than the accelerations—that are associated with injury risk, an elevated PTA under these conditions does not inherently imply increased mechanical load on the body.

A similar mechanism may help explaining the findings of the present study. In the present study, we observed increased PTA, evVel, and PAV during the later stages of the half-marathon run, while SF, strLen, and FSA remained unchanged. These findings suggest that muscular fatigue may have contributed to both a reduction in effective mass—through subtle kinematic changes such as increased knee flexion—and to less controlled and faster foot rollover movements in both the frontal (evVel) and sagittal planes (PAV). The combination of a reduced effective mass and increased segmental angular velocities likely led to the observed increase in PTA, despite the absence of changes in both strLen and SF.

In contrast to strLen and SF, both t_C increased and t_F decreased significantly over the course of the run, leading to a significant rise in DF. DF represents the proportion of the gait cycle spent in ground contact (Equation (2)). At constant running speed, a higher DF suggests a shift toward a more compliant or cautious running pattern, characterized by lower leg stiffness, increased knee flexion at IC, and vertical displacement, and reduced vertical ground reaction forces [11,53]. Additionally, muscular fatigue—particularly in the plantarflexors and hip extensors—could delay the push-off phase, contributing to a longer stance duration [54]. This prolonged t_C may serve as a stabilizing strategy in the presence of reduced neuromuscular control, but may also facilitate the transmission of higher impact accelerations, as observed in the increased PTA. Similar observations have been reported by Möhler et al. [55] and Giovanelli et al. [19]: In both studies, SF remained constant, while t_C increased and t_F decreased, resulting in a higher DF. In the study by Möhler et al. [55], t_C increased significantly, while t_F decreased. Likewise, Giovanelli et al. [19] found a comparable shift during a mountain run, with t_C increasing and t_F decreasing, despite constant running speed and SF.

Based on our findings regarding the spatiotemporal parameters, Hypothesis (IV) can also only be partially accepted.

4.5. Limitations and Recommendations

This study has a few limitations. Firstly, the generalizability of our findings is limited by the study population. Since participants were primarily recreational runners with speeds between 10 and 13 km/h, results may not apply to elite athletes or slower beginners. The broad variability in fitness levels likely increased variance and reduced effect sizes. Future studies may benefit from recruiting more homogeneous subgroups (e.g., novice or experienced runners) to reduce variability. Additionally, instead of self-selected speeds, future protocols could include an incremental test prior to data collection, allowing testing at a defined relative intensity (e.g., near the anaerobic threshold) to ensure comparable loading across participants and enhance validity [9,34,40]. Secondly, one limitation of the measurement system was cable breakage in some participants due to mechanical stress on the connections, which restricted analysis to one leg. Future studies should consider using more robust cables or potentially wireless systems to allow for bilateral data collection and the assessment of interlimb coordination. However, any wireless system must provide sufficiently high sampling frequency to ensure accurate data acquisition [46]. Thirdly, the study by Derrick et al. [11] demonstrated that muscular fatigue is associated with increased knee flexion. We adopted this assumption. In future research, an IMU on each thigh could enable measurement of knee joint angles and investigation of fatigue effects on knee kinematics during prolonged running under field conditions.

5. Conclusions

This study highlights the potential of IMU-based systems for continuous, field-based monitoring of running biomechanics under real-world fatigue conditions. Unlike previous studies limited by pre–post comparisons, laboratory constraints, or assessed only a narrow range of variables, our findings provide ecologically valid insights into the biomechanical adaptations occurring during prolonged endurance runs. The results demonstrate that muscular fatigue in the latter stages of a half-marathon is associated with increased PTA, PAV, and evVel, while FSA, SF and strLen remained unchanged. These changes likely reflect a reduction in effective mass and less controlled segmental motion. The observed increase in DF may indicate a fatigue-related adaptation through prolonged t_C, potentially linked to reduced leg stiffness and delayed push-off.

From a practical perspective, these findings have several important implications for athletes, coaches, clinicians, and physiotherapists. The continuous monitoring of PTA, PAV, and evVel during real-time running opens new opportunities for early detection of fatigue-related deterioration in movement quality. This enables timely interventions—such as gait adjustments or terminating a training session—to prevent overload. Furthermore, the observed increases in segmental angular velocities underline the importance of targeted training for distal lower limb muscles to maintain control under fatigue. Strength and plyometric training aimed at preserving leg stiffness and reducing t_C may further enhance fatigue resistance. Additionally, bilateral data collection offers valuable insight into asymmetric loading patterns, which may signal compensatory mechanisms or emerging imbalances. Early identification of such patterns can contribute to individualized training strategies and effective injury prevention.

Overall, this study highlights the practical value of IMU technology for analyzing fatigue-related biomechanical changes and preventing running-related overuse injuries.

Author Contributions: Conceptualization, P.K., C.M. and T.L.M.; methodology, P.K. and C.M.; software, C.M.; validation, P.K. and C.M.; formal analysis, P.K., C.M. and T.H.; investigation, P.K.; resources, P.K. and C.M.; data curation, P.K. and C.M.; writing—original draft preparation, C.M.; writing—review and editing, P.K., C.M., T.L.M. and T.H.; visualization, T.H.; supervision, P.K.; project administration, P.K., C.M. and T.L.M.; funding acquisition, T.L.M. and C.M. All authors have read and agreed to the published version of the manuscript.

Funding: This research received no external funding.

Institutional Review Board Statement: All procedures were conducted according to the recommendations of the Declaration of Helsinki and were approved by the Ethics Committee of Chemnitz University of Technology (reference number #101627696).

Informed Consent Statement: Informed consent was obtained from all subjects involved in the study.

Data Availability Statement: The dataset used and analyzed in this study is available from the corresponding author upon reasonable request.

Acknowledgments: The authors thank the runners for their dedication to this study.

Conflicts of Interest: The authors declare no conflicts of interest.

Abbreviations

The following abbreviations are used in this manuscript:

PTA	Peak tibial acceleration
evVel	Peak eversion velocity
FSA	Foot strike angle
PAV	Peak angular velocity in the sagittal plane
SF	Stride frequency
strLen	Stride length
DF	Duty Factor
t_C	Ground contact time
t_F	Flight time
IC	Initial ground contact
HR	Heart Rate
RPE	Rating of perceived exertion
IMU	Inertial measurement unit
PFP	Patellofemoral pain
MTSS	Medial tibial stress syndrome
TSF	Tibial stress fractures
bpm	Beats per minute

References

1. Hennig, E.M. Eighteen years of running shoe testing in Germany—A series of biomechanical studies. *Footwear Sci.* **2011**, *3*, 71–
[CrossRef]
2. Yang, S.-J.; Yang, F.; Gao, Y.; Su, Y.-F.; Sun, W.; Jia, S.-W.; Wang, Y.; Lam, W.-K. Gender and Age Differences in Performance Over 70,000 Chinese Finishers in the Half- and Full-Marathon Events. *Int. J. Environ. Res. Public Health* **2022**, *19*, 7802. [CrossRe
3. Anthony, D.; Rüst, C.; Cribari, M.; Rosemann, T.; Lepers, R.; Knechtle, B. Differences in participation and performance trends age group half and full marathoners. *Chin. J. Physiol.* **2014**, *57*, 209–219. [CrossRef] [PubMed]
4. Van Gent, R.N.; Siem, D.; van Middelkoop, M.; van Os, A.G.; Bierma-Zeinstra, S.M.A.; Koes, B.W. Incidence and determinants lower extremity running injuries in long distance runners: A systematic review. *Br. J. Sports Med.* **2007**, *41*, 469–480, discussi 480. [CrossRef]
5. Ceyssens, L.; Vanelderen, R.; Barton, C.; Malliaras, P.; Dingenen, B. Biomechanical Risk Factors Associated with Running-Relat Injuries: A Systematic Review. *Sports Med.* **2019**, *49*, 1095–1115. [CrossRef]
6. Fredette, A.; Roy, J.-S.; Perreault, K.; Dupuis, F.; Napier, C.; Esculier, J.-F. The Association Between Running Injuries and Traini Parameters: A Systematic Review. *J. Athl. Train.* **2022**, *57*, 650–671. [CrossRef] [PubMed]
7. Camelio, K.; Gruber, A.H.; Powell, D.W.; Paquette, M.R. Influence of Prolonged Running and Training on Tibial Acceleration ar Movement Quality in Novice Runners. *J. Athl. Train.* **2020**, *55*, 1292–1299. [CrossRef]
8. Verbitsky, O.; Mizrahi, J.; Voloshin, A.; Treiger, J.; Isakov, E. Shock Transmission and Fatigue in Human Running. *J. Appl. Biome* **1998**, *14*, 300–311. [CrossRef]
9. Mizrahi, J.; Verbitsky, O.; Isakov, E.; Daily, D. Effect of fatigue on leg kinematics and impact acceleration in long distance runni *Hum. Mov. Sci.* **2000**, *19*, 139–151. [CrossRef]
10. Dierks, T.A.; Manal, K.T.; Hamill, J.; Davis, I. Lower extremity kinematics in runners with patellofemoral pain during a prolong run. *Med. Sci. Sports Exerc.* **2011**, *43*, 693–700. [CrossRef]
11. Derrick, T.R.; Dereu, D.; McLean, S.P. Impacts and kinematic adjustments during an exhaustive run. *Med. Sci. Sports Exerc.* **20** *34*, 998–1002. [CrossRef]
12. Giandolini, M.; Gimenez, P.; Temesi, J.; Arnal, P.J.; Martin, V.; Rupp, T.; Morin, J.-B.; Samozino, P.; Millet, G.Y. Effect of the Fatig Induced by a 110-km Ultramarathon on Tibial Impact Acceleration and Lower Leg Kinematics. *PLoS ONE* **2016**, *11*, e01516 [CrossRef]

13. Reenalda, J.; Maartens, E.; Homan, L.; Buurke, J.H.J. Continuous three dimensional analysis of running mechanics during a marathon by means of inertial magnetic measurement units to objectify changes in running mechanics. *J. Biomech.* **2016**, *49*, 3362–3367. [CrossRef] [PubMed]
14. Chan-Roper, M.; Hunter, I.; W Myrer, J.; L Eggett, D.; K Seeley, M. Kinematic changes during a marathon for fast and slow runners. *J. Sports Sci. Med.* **2012**, *11*, 77–82. [PubMed]
15. Chen, T.L.-W.; Wong, D.W.-C.; Wang, Y.; Tan, Q.; Lam, W.-K.; Zhang, M. Changes in segment coordination variability and the impacts of the lower limb across running mileages in half marathons: Implications for running injuries. *J. Sport Health Sci.* **2022**, *11*, 67–74. [CrossRef] [PubMed]
16. Hardin, E.C.; van den Bogert, A.J.; Hamill, J. Kinematic adaptations during running: Effects of footwear, surface, and duration. *Med. Sci. Sports Exerc.* **2004**, *36*, 838–844. [CrossRef]
17. Morin, J.-B.; Samozino, P.; Millet, G.Y. Changes in running kinematics, kinetics, and spring-mass behavior over a 24-h run. *Med. Sci. Sports Exerc.* **2011**, *43*, 829–836. [CrossRef]
18. Morio, C.Y.-M.; Garcia, S.; Flores, N. Effects of one-hour road running and shoe on tibial accelerations in recreational runners. *ISBS Proc. Arch.* **2020**, *38*, 144–147.
19. Giovanelli, N.; Taboga, P.; Rejc, E.; Simunic, B.; Antonutto, G.; Lazzer, S. Effects of an Uphill Marathon on Running Mechanics and Lower-Limb Muscle Fatigue. *Int. J. Sports Physiol. Perform.* **2016**, *11*, 522–529. [CrossRef]
20. Prigent, G.; Apte, S.; Paraschiv-Ionescu, A.; Besson, C.; Gremeaux, V.; Aminian, K. Concurrent Evolution of Biomechanical and Physiological Parameters With Running-Induced Acute Fatigue. *Front. Physiol.* **2022**, *13*, 814172. [CrossRef]
21. Olaya-Cuartero, J.; Pueo, B.; Villalon-Gasch, L.; Jimenez-Olmedo, J.M. Stability of Running Stride Biomechanical Parameters during Half-Marathon Race. *Appl. Sci.* **2024**, *14*, 4807. [CrossRef]
22. Ruder, M.; Jamison, S.T.; Tenforde, A.; Mulloy, F.; Davis, I.S. Relationship of Foot Strike Pattern and Landing Impacts during a Marathon. *Med. Sci. Sports Exerc.* **2019**, *51*, 2073–2079. [CrossRef]
23. DeJong Lempke, A.F.; Hunt, D.L.; Willwerth, S.B.; d'Hemecourt, P.A.; Meehan, W.P.; Whitney, K.E. Biomechanical changes identified during a marathon race among high-school aged runners. *Gait Posture* **2024**, *108*, 44–49. [CrossRef] [PubMed]
24. Hennig, E.M.; Milani, T.L.; Lafortune, M.A. Use of Ground Reaction Force Parameters in Predicting Peak Tibial Accelerations in Running. *J. Appl. Biomech.* **1993**, *9*, 306–314. [CrossRef] [PubMed]
25. Mercer, J.A.; Bates, B.T.; Dufek, J.S.; Hreljac, A. Characteristics of shock attenuation during fatigued running. *J. Sports Sci.* **2003**, *21*, 911–919. [CrossRef] [PubMed]
26. Davis, I.; Milner, C.; Hamill, J. Does increased loading during running lead to tibial stress fractures? A prospective study. *Med. Sci. Sports Exerc.* **2004**, *36*, S58.
27. Pohl, M.B.; Mullineaux, D.R.; Milner, C.E.; Hamill, J.; Davis, I.S. Biomechanical predictors of retrospective tibial stress fractures in runners. *J. Biomech.* **2008**, *41*, 1160–1165. [CrossRef]
28. Stacoff, A.; Reinschmidt, C.; Nigg, B.M.; van den Bogert, A.J.; Lundberg, A.; Denoth, J.; Stüssi, E. Effects of foot orthoses on skeletal motion during running. *Clin. Biomech.* **2000**, *15*, 54–64. [CrossRef]
29. Vtasalo, J.T.; Kvist, M. Some biomechanical aspects of the foot and ankle in athletes with and without shin splints. *Am. J. Sports Med.* **1983**, *11*, 125–130. [CrossRef]
30. Milner, C.E.; Ferber, R.; Pollard, C.D.; Hamill, J.; Davis, I.S. Biomechanical factors associated with tibial stress fracture in female runners. *Med. Sci. Sports Exerc.* **2006**, *38*, 323–328. [CrossRef]
31. Shih, Y.; Ho, C.-S.; Shiang, T.-Y. Measuring kinematic changes of the foot using a gyro sensor during intense running. *J. Sports Sci.* **2014**, *32*, 550–556. [CrossRef]
32. Hazzaa Walaa Eldin, A.; Mattes, K. The effect of local muscle fatigue and foot strike pattern during barefoot running at different speeds. *Dtsch. Z. Sport.* **2019**, *70*, 175–182. [CrossRef]
33. Bräuer, S.; Kiesewetter, P.; Milani, T.L.; Mitschke, C. The 'Ride' Feeling during Running under Field Conditions-Objectified with a Single Inertial Measurement Unit. *Sensors* **2021**, *21*, 5010. [CrossRef]
34. Mizrahi, J.; Verbitsky, O.; Isakov, E. Fatigue-related loading imbalance on the shank in running: A possible factor in stress fractures. *Ann. Biomed. Eng.* **2000**, *28*, 463–469. [CrossRef]
35. Milner, C.E.; Hawkins, J.L.; Aubol, K.G. Tibial Acceleration during Running Is Higher in Field Testing Than Indoor Testing. *Med. Sci. Sports Exerc.* **2020**, *52*, 1361–1366. [CrossRef]
36. Hill, M.; Kiesewetter, P.; Milani, T.L.; Mitschke, C. An Investigation of Running Kinematics with Recovered Anterior Cruciate Ligament Reconstruction on a Treadmill and In-Field Using Inertial Measurement Units: A Preliminary Study. *Bioengineering* **2024**, *11*, 404. [CrossRef] [PubMed]
37. Bruneau, M.M.; Heiderscheit, B.C.; Morgan, K.D.; Moore, T.E.; Denegar, C.R.; Casa, D.J.; Devaney, L.L. Comparison of Step Rate and Tibial Acceleration During Treadmill and Real-world Running. *JOSPT Open* **2025**, *3*, 77–84. [CrossRef]

38. Milner, C.E.; Foch, E.; Gonzales, J.M.; Petersen, D. Biomechanics associated with tibial stress fracture in runners: A systemati review and meta-analysis. *J. Sport Health Sci.* **2023**, *12*, 333–342. [CrossRef] [PubMed]
39. Xiang, L.; Gao, Z.; Wang, A.; Shim, V.; Fekete, G.; Gu, Y.; Fernandez, J. Rethinking running biomechanics: A critical revie of ground reaction forces, tibial bone loading, and the role of wearable sensors. *Front. Bioeng. Biotechnol.* **2024**, *12*, 137738 [CrossRef]
40. Stirling, L.M.; von Tscharner, V.; Fletcher, J.R.; Nigg, B.M. Quantification of the manifestations of fatigue during treadmill runnin *Eur. J. Sport Sci.* **2012**, *12*, 418–424. [CrossRef]
41. Kiesewetter, P.; Bräuer, S.; Haase, R.; Nitzsche, N.; Mitschke, C.; Milani, T.L. Do Carbon-Plated Running Shoes with Differen Characteristics Influence Physiological and Biomechanical Variables during a 10 km Treadmill Run? *Appl. Sci.* **2022**, *12*, 794 [CrossRef]
42. Mitschke, C.; Heß, T.; Milani, T. Which Method Detects Foot Strike in Rearfoot and Forefoot Runners Accurately when Using a Inertial Measurement Unit? *Appl. Sci.* **2017**, *7*, 959. [CrossRef]
43. Sabatini, A.M.; Martelloni, C.; Scapellato, S.; Cavallo, F. Assessment of walking features from foot inertial sensing. *IEEE Tran Biomed. Eng.* **2005**, *52*, 486–494. [CrossRef] [PubMed]
44. Alexander, R.M. Energy-saving mechanisms in walking and running. *J. Exp. Biol.* **1991**, *160*, 55–69. [CrossRef]
45. Mitschke, C.; Öhmichen, M.; Milani, T. A Single Gyroscope Can Be Used to Accurately Determine Peak Eversion Velocity durin Locomotion at Different Speeds and in Various Shoes. *Appl. Sci.* **2017**, *7*, 659. [CrossRef]
46. Mitschke, C.; Zaumseil, F.; Milani, T.L. The influence of inertial sensor sampling frequency on the accuracy of measuremen parameters in rearfoot running. *Comput. Methods Biomech. Biomed. Eng.* **2017**, *20*, 1502–1511. [CrossRef]
47. Cohen, J. A Power Primer. *Psychol. Bull.* **1992**, *112*, 155–159. [CrossRef]
48. Olaya-Cuartero, J.; Lopez-Arbues, B.; Jimenezolmedo, J.M.; Villalon-Gasch, L. Influence of Fatigue on the Modification o Biomechanical Parameters in Endurance Running: A Systematic Review. *Int. J. Exerc. Sci.* **2024**, *17*, 1377–1391. [CrossRef]
49. Hreljac, A. Impact and overuse injuries in runners. *Med. Sci. Sports Exerc.* **2004**, *36*, 845–849. [CrossRef]
50. Heß, T.; Milani, T.L.; Stoll, J.; Mitschke, C. Running and Jumping After Muscle Fatigue in Subjects with a History of Knee Injury What Are the Acute Effects of Wearing a Knee Brace on Biomechanics? *Bioengineering* **2025**, *12*, 661. [CrossRef]
51. Breine, B.; Malcolm, P.; Galle, S.; Fiers, P.; Frederick, E.C.; de Clercq, D. Running speed-induced changes in foot contact patter influence impact loading rate. *Eur. J. Sport Sci.* **2019**, *19*, 774–783. [CrossRef]
52. Heidenfelder, J.; Sterzing, T.; Bullmann, M.; Milani, T.L. Heel strike angle and foot angular velocity in the sagittal plane durin running in different shoe conditions. *J. Foot Ankle Res.* **2008**, *1*, O16. [CrossRef]
53. Hanley, B.; Tucker, C.B.; Gallagher, L.; Parelkar, P.; Thomas, L.; Crespo, R.; Price, R.J. Grizzlies and gazelles: Duty factor is a effective measure for categorizing running style in English Premier League soccer players. *Front. Sports Act. Living* **2022**, *4*, 93967 [CrossRef]
54. Apte, S.; Prigent, G.; Stöggl, T.; Martínez, A.; Snyder, C.; Gremeaux-Bader, V.; Aminian, K. Biomechanical Response of the Lowe Extremity to Running-Induced Acute Fatigue: A Systematic Review. *Front. Physiol.* **2021**, *12*, 646042. [CrossRef]
55. Möhler, F.; Fadillioglu, C.; Stein, T. Fatigue-Related Changes in Spatiotemporal Parameters, Joint Kinematics and Leg Stiffness i Expert Runners During a Middle-Distance Run. *Front. Sports Act. Living* **2021**, *3*, 634258. [CrossRef]

MDPI AG
Grosspeteranlage 5
4052 Basel
Switzerland
Tel.: +41 61 683 77 34

Biomechanics Editorial Office
E-mail: biomechanics@mdpi.com
www.mdpi.com/journal/biomechanics

www.ingramcontent.com/pod-product-compliance
Lightning Source LLC
Chambersburg PA
CBHW062040220926
51850CB00010B/2614
9783725869343